Genesis and the Chemical Philosophy

AMS STUDIES IN THE RENAISSANCE
ISSN 0195-8011
Series ISBN-13: 978-0-404-61460-7

No. 45

GENESIS AND THE CHEMICAL PHILOSOPHY

True Christian Science in the Sixteenth and Seventeenth Centuries

ISBN-13: 978-0-404-62345-6

Genesis and the Chemical Philosophy

True Christian Science in the Sixteenth and Seventeenth Centuries

Michael T. Walton

AMS Press, Inc.
New York

Library of Congress Cataloging-in-Publication Data

Walton, Michael Thomson, 1945–
Genesis and the chemical philosophy : true Christian science in the sixteenth and seventeenth centuries / Michael T. Walton.
p. cm. — (AMS studies in the Renaissance, ISSN 0195-8011 ; no. 45)
Includes bibliographical references and index.
ISBN 978-0-404-62345-6 (cloth : alk. paper)
1. Science, Renaissance. 2. Alchemy—History. 3. Chemistry—Philosophy—History—16th century. 4. Chemistry—Philosophy—History—17th century. 5. Religion and science—History—16th century. 6. Religion and science—History—17th century. I. Title.
Q125.2.W36 2011
509.4'09031–dc22

2010049135

All AMS books are printed on acid-free paper that meets the guidelines for performance and durability of the Committee on Production Guidelines for Book Longevity of the Council on Library Resources.

AMS Press, Inc.
Brooklyn Navy Yard, 63 Flushing Ave.–Unit #221
Brooklyn, New York 11205-1073, USA
www.amspressinc.com

MANUFACTURED IN THE UNITED STATES OF AMERICA

To my children, Judith, Sarah, Keziah, Rachel A"H, Adam, and Nathan, and to my wife, Phyllis, who is indispensable to my life and research.

CONTENTS

LIST OF FIGURES

All figures are drawn from the author's collection.

PREFACE

Modern approaches to the Pentateuch, such as those that divide it into small text units and attempt to establish its development through history, would have been rejected in the early modern period as heretical.[1] Moshe Goshen-Gottstein has connected the early modern rejection of the "fourfold" biblical exegesis of the Middle Ages in favor of a philological inquiry into the "literal sense" with the rise of modern biblical scholarship in the nineteenth century.[2] However, early modern biblical philologers and exegetes, like their scholastic forerunners, were committed to the truth and inerrancy of the Pentateuch. They wanted to understand the text in order to know God's will. In this they differed little from the early Christians who conquered the pagan Roman Empire and established the religious writings of Jews, a small, obstreperous,

1 The sources, methodologies, and complexities of modern biblical study are presented in a collection of articles in *Qumran and the History of the Biblical Text*, ed. Frank M. Cross and Shemaryahu Talmon (Cambridge, MA: Harvard University Press, 1975). In that collection, of especial interest for the history of the study of the biblical text is the article by Moshe H. Goshen-Gottstein, "Hebrew Biblical Manuscripts: Their History and Their Place in the HUBP Edition," 42–89.

2 Moshe H. Goshen-Gottstein, "Foundations of Biblical Philology in the Seventeenth Century: Christian and Jewish Dimensions," in *Jewish Thought in the Seventeenth Century*, ed. Isadore Twersky and Bernard Septimus (Cambridge, MA: Harvard University Press, 1987), 77–94.

but significant religious group within and without the Empire, as the true guide for God's new chosen people.

The centrality of the Pentateuch and the rest of the Jewish scriptures to the Christian religion made them essential to the Christian use of pagan knowledge. Pagan texts and ideas came to be tested and interpreted against a Jewish background. Without the rise of Christianity, it is doubtful that the ideas, ideals, and especially the cosmogony of the Jews would have come to dominate western religious thought. The biblical cosmogony became a major focus in the integration of pagan with Jewish ideas. That integration became the foundation of late classical and medieval Christianity. Patristic fathers and medieval doctors supplemented Genesis with Plato, Aristotle, Ptolemy, and others in order to establish a Christian understanding of the creation that combined theological, philosophical, and "natural" truths.

The process of Judeo-Christian-pagan integration also took place within the ancient mystical-alchemical tradition. Technical chemical-artisan texts and gnostic-mystical texts were read in light of and overlaid with Genesis and Christian theology. Christianized alchemy continued into the Middle Ages and Renaissance. Like biblical exegesis, Christian alchemy underwent a reinterpretation, if not a reformation, in the early modern period, becoming the "chemical philosophy," the subject of this book. I will use the term "chemical philosophy" in a restricted sense as the conjunction of alchemical, Neoplatonic/Neo-Pythagorean, Kabbalistic ideas developed by Paracelsus and like-minded natural philosophers. I exclude those who used chemical notions, such as Guinter of Andernach, but were not true Paracelsians. I also exclude many of those called Paracelsians as an epithet by their opponents. The name "Paracelsian," like "Judaizer," was so often used during polemical exchanges that it lost much of its meaning.[3] Following the example of Luther, who by studying the scriptures had cleansed religion of idolatry, Paracelsus wanted to purify the study of nature from the paganism that had long infected it. He believed that the book of nature could be read properly only in light of the true understanding of the book of

---

3 There were chemists and theologians, such as Andreas Libavius, Germani Courtin, and Thomas Erastus, who disliked the mysticism and heterodox religiosity of Paracelsus and his disciples. The ideas of the anti-Paracelsian chemists, for the most part, are peripheral to the importance of Genesis in the chemical philosophy. They were, however, important in the development of chemistry. See Allen G. Debus, *The Chemical Philosophy* (New York: Science History Publications, 1977), 127–204.

scripture. Paracelsus, and the chemists he inspired, wanted to understand the forces of nature and use them to benefit humanity. They "knew" that true chemical science would agree with Genesis, which was God's unambiguous revelation to mankind.

Paracelsus's immediate successors followed his ideas closely in their systematizing of his philosophy. Gerhard Dorn, Oswald Croll, and Peter Severinus were true advocates of the master. They adhered to his theories, terminology, and deep commitment to Genesis as the point of departure for correct chemical thought. Later chemical philosophers, such as Heinrich Khunrath and Robert Fludd, were much more idiosyncratic in their philosophies. They were profound scholars with baroque sensitivities, for whom Paracelsus was an important, but not the only, philosophical-chemical point of reference. Other chemists, such as Joan Baptista Van Helmont, Johann Rudolph Glauber, and Johann Joachim Becher, were critical of Paracelsus. They were practiced laboratory workers whose understanding of chemistry was informed by experience and experiments, as well as philosophy. Khunrath, Fludd, Van Helmont, Glauber, and Becher, like Paracelsus and the earlier chemical philosophers, looked to the book of scripture in general and Genesis in particular as the source of chemical and, indeed, all natural philosophical knowledge.

In the following pages, I will outline the complex intellectual traditions surrounding the interaction of chemistry and Genesis from classical times into the seventeenth century. I will detail the baptism of chemistry into a Christian natural philosophy by Paracelsus and his heirs in the sixteenth and seventeenth centuries. Chemical philosophers reexamined matter theory in light of Genesis. They developed a new epistemology, which focused on experiencing nature rather than relying on accepted texts. This attitude fostered quantitative experimentation, which ultimately transformed chemistry. With this transformation, Genesis itself lost its importance; the "reading" of nature was no longer dependent on theological considerations. Chemistry moved from a theological to a secular interpretation of nature, as is found in modern science.

Acknowledgment

I thank my teacher and friend Allen G. Debus and my colleague and friend Raymond B. Waddington for their wisdom and editorial skill.

1. In the Beginning . . . : The Theological Background

The creation account found in Genesis is crucial not only to western theology but also to western natural thought. The conflict between science and religion in the nineteenth and twentieth centuries tends to obscure the long period of harmony between the two from the ancient world until the early modern discomfiture of Galileo and the nineteenth-century "crisis of Darwin."[1] Because divine revelation was true, the theologian and the natural philosopher (often the same person) were constrained to shape and fit their ideas to the outline provided by holy writ. The poetic beauty of the first lines of Genesis

---

1 The present study will demonstrate that the chemical philosophy begun by Paracelsus was very much a part of the Christian world of the early modern period. Indeed, western science in general arose in a Christian theological-philosophical context. When theological authorities reacted to Galileo and other scientists, they demonstrated that the methodologies of the various branches of natural philosophy yielded results at variance with accepted truths. A profoundly religious chemist, Robert Boyle acknowledged the different universes of discourse in science and religion when he declared in the *The Sceptical Chymist* that he should speak as a natural philosopher and not as a philologer. From the perspective of what the present study sets out to show, science, in its manifestation as chemistry, moved beyond Christian theology by developing an experimental methodology, a methodology that made it independent of theology. These points will be detailed in chapter 6. Religious fundamentalists in the last quarter of the twentieth century attempted to reunite science and theology in "creation science." This movement will be discussed in the epilogue.

contained neither homily nor natural speculation. Nonetheless, the terse pronouncements of those few verses occasioned great diversity of thought and interpretation. Indeed, within natural philosophy, Genesis easily fit, or was made to fit, various concepts of matter and natural processes. In a way, the majesty of the first ten verses of the Bible formed a bridge between theology and nature:

> In the beginning God created the heaven and the earth. And the earth was without form, and void; and the darkness was upon the face of the deep. And the Spirit of God moved upon the face of the waters.[2] And God said, Let there be light: and there was light. And God saw the light, that it was good: and God divided the light from the darkness. And God called the light Day, and the darkness he called Night. And the evening and the morning were the first day.
>
> And God said, Let there be a firmament in the midst of the waters, and let it divide the waters from the waters. And God made the firmament, and divided the waters which were under the firmament from the waters which were above the firmament: and it was so. And God called the firmament Heaven. And the evening and the morning were the second day.
>
> And God said, Let the waters under the heaven be gathered together unto one place, and let the dry land appear: and it was so. And God called the dry land Earth; and the gathering together of the waters he called Seas: and God saw that it was good.
>
> (King James Version, 1611)

Genesis bridged theology and natural thought because of its openness to interpretation, arising from a poetic ambiguity of the concept of Chaos, from which the elements, including Light and Water, emerged. Creation myths, including those that dealt with concepts of matter, were popular in the ancient world. Those philosophies that dispensed with the anthropomorphic actions of particular deities and focused on the behavior of matter and form fostered ideas that were not necessarily antithetical to Judaism or Christianity.

---

[2] This phrase became crucial for sixteenth- and seventeenth-century chemists.

Indeed, Jewish and Christian biblical commentators noted the similarities between pagan and biblical ideas of creation and matter.

There is broad agreement among scholars that what was to become the Masoretic text of the Pentateuch was established by the first century BCE. The Qumran manuscripts of Genesis show only minor variations from the *textus receptus*. The Qumran materials further indicate that the received text is part of a Babylonian textual tradition while the Greek text of the Septuagint is related to a Palestinian textual tradition.[3] The received text and the Septuagint composed the book of Genesis available to and used by Jews, early Christians, and their leaders. The Gospels relied on the Septuagint. Jerome's translation into Latin made use of both the Hebrew and Greek Bibles.

Christian biblical exegesis began in the first century CE. As Christianity rapidly became a gentile movement, its exegesis came to encompass both its Jewish roots and its Greco-Roman milieu.[4] This syncretism was manifest in discussions of the Genesis text. Christians and Jews used pagan natural philosophy to explicate the text as part of a syncretistic approach to Genesis.[5] The doctrine of the four elements and notions of the formation of minerals and metals, however, were used in a predominately theological context. They did not control the direction of commentaries, but rather added details to them. Only the alchemical texts of the ancient and medieval worlds and, later, the "chemical writings" of the sixteenth and seventeenth centuries treated Genesis as primarily a natural philosophical work and *the* key to understanding nature.

---

3 For a discussion of the history of the biblical text, see Frank Moore Cross, "The History of the Biblical Text in Light of Discoveries in the Judean Desert," in *Qumran and the History of the Biblical Text*, ed. Cross and Talmon, 177–95, and Frank Moore Cross, *From Epic to Canon* (Baltimore, MD: Johns Hopkins University Press, 1998). The Jewish and Christian exegetical traditions, including the works of Philo mentioned below, are briefly but insightfully discussed by James L. Kugel and Rowen A. Greer in *Early Biblical Interpretation* (Philadelphia: Westminster Press, 1986).

4 There is a large body of literature on gentile Christianity. Three classic texts are A. D. Nock, *Early Gentile Christianity and Its Hellenistic Background* (New York: Harper, 1964); James Parker, *The Conflict of the Church and the Synagogue* (New York: Atheneum, 1979); and E. P. Sanders, *Paul and Palestinian Judaism* (Philadelphia: Fortress Press, 1977).

5 In Talmud Bavli, *Sota* 49b, it is stated that R. Gamliel had 500 students who studied Hebrew [law] and another 500 who studied Greek [wisdom]. In Talmud Bavli, *Pesachim* 94b, R. Judah the Prince said that the sages of the nations' view on the movement of the sun was preferable to that of the rabbis. The Talmudic sages were very conscious of pagan learning.

I

Jewish tradition taught and teaches that the Torah is written in the language of humans in order that they can comprehend it.[6] It follows that Genesis was written in Hebrew terms that were intelligible to the oriental mind of the Jew. It does not partake of Greek rationalism. Rather, Genesis uses words and concepts drawn from the Hebrew culture as well as surrounding near eastern societies. For example, the Hebrew word for death is *mêt*; the Canaanite god of death is *Môt.* At this basic level of vocabulary, Genesis connects to ancient near eastern culture as a whole. On a macro level, there is a similar convergence. The first verses of Genesis mention chaos, light, darkness, and water. All of these were names for both gods and the visible building blocks of the world according to Sumerian, Babylonian, and Persian cosmologies (e.g., the water monster Tiamat, the god of light Ahura Mazda, and the god of darkness Ahriman). Genesis says that these aspects of nature are not gods, or even powers in and of themselves, but rather creations of the one God. Genesis, at this level, transforms ancient near eastern lore into a new cosmogony that refutes polytheism. The Genesis text itself begins a process of harmonizing monotheism with ancient near eastern non-monotheistic ideas about the created world.

The idea that the world arose from chaos was not uniquely Hebraic or even original with the Jews. The Babylonian creation myth recounted how Marduk slew and divided the water monster Tiamat. The Phoenician creation myth described an original windy mist and turbid chaos. Early Greek philosophers taught that *ater* was boundless and contained the principles of condensation and rarefaction. These concepts fit with the creation story in Genesis, but also with what may be earlier texts, such as Psalm 24:2, "For it is He who has founded it [the world] upon the seas and established it on the floods," and Isaiah 42:5, "[God] creates the heavens and stretches them out, spreads out the earth. . . ."

The transmission of a text, especially a divine text such as Genesis, is an interpretive activity. The transmitters decide what is worth copying for their own use and for the use of others. They often insert comments in order to explain parts of the transmitted texts that may have become unclear or may seem to require elaboration. This type of "commentary" is present in the

---

6 Talmud Bavli, *Yevamoth* 71a and *Bava metzia* 31b.

Pentateuch with its notations that "the Canaanites were then in the land" (Gen. 12:6) and that you will make a fringe on your garments, "that you may look upon it, and remember all the commandments" (Num. 15:39). A more overt type of commentary involves translation. In the Greco-Roman world, the public reading of the Pentateuch in Hebrew was accompanied by a running interpretive translation, a Targum. The most important Aramaic Targums are ascribed to Onkelos and Jonathan and the most famous Greek Targum to Aquila. The interpretive nature of the Targum is seen in Onkelos's rendering of Genesis 1:2. The Hebrew text reads, "The Spirit of God moved on the face of the waters." Onkelos states in a naturalistic vein, "The Lord caused an east wind to blow on the face of the waters."[7] Onkelos's approach to the text is consistent with the Talmud's statement that "The world follows its customary order."[8]

A great challenge for Jews and Christians of the ancient world was to harmonize the Hebrew scriptures and especially Genesis with the Greco-Roman ideas of nature and its processes. As a unique, divine, and true revelation, the Hebrew scriptures took precedence over any human wisdom. Still, God had endowed humans with the ability to reason and to understand and the Greeks and Romans had made obvious progress in comprehending nature. How did the light of natural reason comport with revealed truth? This question greatly occupied Jewish and Christian believers in the late classical world. Indeed, from Philo of Alexandria and the church fathers to the present, there is a remarkable continuity of thought and approaches to harmonizing secular and religious knowledge.

Commentators developed two basic approaches to pagan science and Genesis. Philo the Jew and, later, Origen the Christian viewed Genesis allegorically and hence were able to interpret it to include most ancient pagan wisdom. Other influential Christians, such as Basil of Caesarea and his brother Gregory of Nyssa, believed that the Bible meant what it said and that, therefore, it had to be taken literally. These literalists accepted certain impediments and limits to assimilating pagan ideas into the holy edifice of revealed religion.

Philo (ca. 20 BCE–50 CE) was a Greek-literate Jew who lived in Alexandria, Egypt, a city with a large Jewish community.[9] Although his writings

---

7 Translation mine.

8 Talmud Bavli, *Avodah zara* 54.

9 There is a vast literature on Philo. The most up-to-date edition of Philo in English is the update of C. D. Yonge's 1854 translation corrected and introduced by David M. Scholer,

seem to have had little or no importance in the development of rabbinical thought, they were very popular among both pagans and Christians. Philo apparently was not fluent in Hebrew, but was well educated in Greek and Greek literature. He relied on the Septuagint for his knowledge of Jewish history. His writings connect the intellectual traditions of the Hellenistic-Roman world to that of the Jews. Philo's syncretism helps explain his influence on early Christian theologians who were, themselves, part of both the pagan world and the newly emerging Christian culture.

In "On the Creation of the World," Philo conceives of the creation as a manifestation of the Divine Intellect informing chaotic matter. Archetypical ideas invest substance with order and harmony. The six days of creation are not periods of time, but a way of describing an arrangement of the world; God needed no time to create the world, but the number six is crucial for understanding his creation:

> III. (13) and [Moses] says that the world was made in six days, not because the Creator stood in need of a length of time (for it is natural that God should do everything at once, but merely by uttering a command, but by even thinking of it): but because the things created required arrangements. . . .[10]

Philo believed that number underlay the structure of the world, creating a cosmic harmony, a belief shared with the Pythagoreans. He continues the passage above by explaining the role of number, especially the number six:

The Works of Philo (Peabody, MA: Hendrickson Publishers, 2002). There is also the twelve-volume Loeb Classical Library *Philo* (Cambridge, MA: Harvard University Press, 1929–53) translated by William Heinemann. Philo's writings reveal much about Jews in the Greek-speaking Roman world and about their understanding of Judaism and its relationship to pagan thought. The Talmud states that the Septuagint is a holy book (like the holy Hebrew Torah scroll, it defiled the hands) and indicates that Jewish prayer and scripture reading in the synagogue could be done in Greek (Talmud Bavli, *Megillah* 8a–9a). Philo was part of that "holy" Jewish-Greek world of the first century CE. (See David T. Runia, "Philo, Alexandrian Jew," in *Exegesis and Philosophy: Studies on Philo of Alexandria* [Aldershot: Variorum, 1990], 1–18.) While his own works apparently did not influence rabbinic thought, Philo's Hellenistic world did influence rabbinic Judaism, as indicated by Greek words and ideas in the Talmud. This was proved by Saul Lieberman (see *Greek in Jewish Palestine* [New York: Jewish Theological Seminary, 1942]). Lieberman argued that R. Oshaiah's ideas in *Genesis Rabbah* 1:1 are similar to Philo's in *De Opificio Mundi*, 16.

10 *De Opificio Mundi*, 3, in Scholer, *Philo*.

> and number is akin to arrangement; and, of all numbers, six is, by the laws of nature, the most productive: for of all the numbers, from the unit upwards, it is the first perfect one, being made equal in its parts, and being made complete by them; the number three being half of it, and the number two a third of it, and the unit a sixth of it, and so to say, it is formed so as to be both male and female. . . .[11]

The male and female principles are crucial to the structure of numerical harmonies. The male-female dynamic is the formative power in the natural world, and the number six

> is made up of the power of both natures; for in existing things the odd number is male, and the even number is female; accordingly, of odd numbers the first is the number three, and of even numbers the first is two, and the two numbers multiplied together make six.[12]

Philo's Neoplatonic formation and Neo-Pythagorean number exegesis connect form and matter through the doctrine of the four elements. The elements, as incorporeal substance, became perceptible through matter: "Then He created the incorporeal substance of water and air."[13] They then became tenable to the senses: "The incorporeal world was then already completed . . . and the world, perceptible by the senses was made on the model of it."[14] The perceptible, corporeal Earth was made fruitful by water mixing with earth. Philo's conception of the creation was compatible with theology and with the accepted doctrines about nature that were central to Greek natural thought and that became the foundation for alchemy. Philo's harmonizing hermeneutic was a pattern for Christians of differing approaches to exegesis, like Origen, who warmed to his allegorical interpretations, and those like St. Basil, who preferred literalism. In their interpretive efforts, advocates of each approach combined seemingly irrefutable pagan ideas with Christian truths.

---

11 Ibid.

12 Ibid.

13 Ibid., 5. Philo's use of the traditional elements is in contrast to Lucretius (ca. ?96–55 BCE), who cast doubt on the four as "primal germs" in favor of atoms, see *De Natura Rerum*, bk. 1. Lucretius's atomism was associated with atheism, so religious commentators disapproved of it.

14 *De Opificio Mundi*, 6, in Scholer, *Philo*.

II

Origen of Alexandria (ca. 185–254) was a towering intellectual figure of the early church, educated in both pagan and Christian knowledge.[15] In a letter to his student Gregory Thaumaturgus, Origen commended the integration of pagan learning with Christianity:

> I should like to see you use all the resources of your mind on Christianity and make that your ultimate object. I hope that to that end you will take from Greek philosophy everything capable of serving as an introduction to Christianity and from geometry and astronomy all ideas useful in expounding the Holy Scriptures; so that what ... philosophers say of geometry, music, grammar, rhetoric, and astronomy, that they assist philosophy, we too may be able to say of philosophy itself in relation to Christianity.[16]

Origen's own acclaimed six-column *Hexapla*, a comparison of the Hebrew and various Greek translations of the scriptures, functioned neither as a commentary nor as a vehicle to integrate pagan and Christian ideas. It was in his theological works, most of which are lost, that Origen evinced the more allegorical, nonliteral approach to scripture that he shared with Philo. In his homily on Genesis 1, Origen emphasizes the spiritual meaning of the creation and the Platonic idea that it took place outside of time:

> According to the letter God calls both the light day and the darkness night. But let us see according to the spiritual meaning why it is that when God, in that beginning which we discussed above, "made heaven and earth," and said, "let there be light" and "divided between the light and the darkness and called the light day and the darkness night," and the text said that "there was evening and there was morning," it did not say: "the first day," but said, "one day." It is because there was not yet time before the world existed. But time begins to exist with the following days. For the

15 See Eusebius, *Ecclesiastical History*, bk 6, chap. 23–36, trans. by Isaac Boyle (Grand Rapids, MI: Baker Book House, 1966), and Lee I. Levine, *Caesarea Under Roman Rule* (Leiden: E. J. Brill, 1975).

16 Origen, *Epistle to Gregory*, 1, translated in Levine, *Caesarea*, 123.

> second day and the third and fourth and all the rest begin to designate time.[17]

St. Basil's interpretation of Genesis was consciously literal, in contrast to that of Origen. His commentary on the six days of creation, given as sermons during Lent, became a staple of Christian thought. These Nine Lenten Sermons expressed his integration of Christian belief and classical doctrines of nature.[18] Basil rejected the Platonic formalism of Philo and Origen and accepted that when Genesis said "water," it meant "water." This more Aristotelian approach was consistent with a naturalistic perception of the creation and the world around us.

Basil set out to prove that God had created the world and to refute the "atheism" of Greek philosophy without rejecting facts or truths that philosophy had revealed about astronomy and the elements. He mused:

> "In the beginning God created the heaven and the earth." I stop struck with admiration at this thought. What shall I first say? Where shall I begin my story? Shall I show forth the vanity of the Gentiles? Shall I exalt the truth of our faith?[19]

Christianity offers a certainty that mere philosophy cannot equal.

> The philosophers of Greece have made much ado to explain nature, and not one of their systems has remained firm [and] unshaken, each being overturned by its successor. It is vain to refute them; they are sufficient in themselves to destroy one

17 Robert E. Heine, *The Fathers of the Church: Origen, Homilies on Genesis and Exodus* (Washington, DC: Catholic University of America Press, 1982), 48.

18 St. Basil (ca. 330–79) received a Greek education in Caesarea, Cappadocia, before being converted to Christianity by Origen's disciple Gregory. He harmonized Christianity with some pagan ideas while distinguishing it from others. The primary English translation of Basil's work is *St. Basil: Letters and Select Works*, trans. with notes by Blomfield Jackson, Vol. 8 in *A Select Library of Nicene and Post-Nicene Fathers*, 2nd ser. (New York: Christian Literature Co., 1895) (alternate title: *The Treatise de Spiritu Sancto, The Nine Homilies of the Hexaemeron nad the Letters of Saint Basil the Great*) (hereafter Jackson, *St. Basil*). Basil's use of Greek science is mentioned by Marshall Clagett, *Greek Science in Antiquity* (New York: Collier Books, 1955), 168–69. See also Frederick W. Norris, "Basil of Caesarea," in *Encyclopedia of Early Christianity*, ed. Everett Ferguson (New York: Garland, 1990), 139.

19 Basil, *Hexaemeron*, 1.2 in Jackson, *St. Basil*, 53.

> another. Those who were too ignorant to rise to a knowledge of a God, could not allow that an intelligent cause presided at the birth of the Universe; a primary error that involved them in sad consequences.[20]

After criticizing atomism, he goes on to approve, in part, Ptolemaic astronomy. Ptolemy (ca. 100–170) based his computational astronomy on three assumptions: the heavens are made of an eternal quintessence and not of the four earthly elements, heavenly bodies are spherical, and celestial motion is circular and uniform (*fig. I*).[21] Basil accepted pagan astronomy's spheres and uniform circular motion but rejected the concept of the eternal, uncreated heavens:

> Do not then imagine, O man! that the visible world is without a beginning; and because the celestial bodies move in a circular course . . . do not believe that bodies impelled by a circular movement are, from their nature, without a beginning.[22]

In like manner, he admitted the four elements into his Genesis theology, for the scriptures, he argued, "compound the elements together." He did not discuss the particulars of elemental theory, which were unimportant to his theological goal:

> [In Genesis,] although there is no mention of the elements, fire, water and air, imagine that they were all compounded together, and you will find water, air and fire, in the earth. For fire leaps out from stone; iron which is dug from the earth produces under friction fire in plentiful measure. . . . The earth contains water, as diggers of wells teach us. It contains air too, as is shown by the vapours that it exhales under the sun's warmth when it is damp.[23]

---

20 Ibid.

21 Ptolemy, *Almagest*, pt. 1, trans. R. Catesby Taliaferro, in *Great Books of the Western World*, vol. 16 (Chicago: Encyclopedia Britannica, 1952).

22 Basil, *Hexaemeron* 1.3 in Jackson, *St. Basil*, 53.

23 Basil, *Hexaemeron* 1.7, *St. Basil*, 56.

DE SACRO BOSCO, CAP. 3. 23

CAPITVLVM TERTIVM
de ortu & occaſu ſignorum, de diuerſitate dierum & noctium, & de diuiſione climatum.

Signorum autem ortus & occaſus dupliciter accipitur, quoniam quantum ad poëtas, & quantum ad aſtronomos. Eſt igitur ortus & occaſus ſignorum, quò

Figure I. Ptolemy's Geocentric World. *The Sphere of Sacro Bosco*. Paris, 1672.

Following pagan astronomers, Basil distinguished between the unchanging heaven and the lower, elemental Earth. The "heaven occupies the higher and the earth the lower position in space, (one sees, in fact, that all which is light ascends towards heaven, and heavy substances fall to the ground),"[24] just as is shown by Aristotle. Basil instructed the Christian, "Do not ask, then, for an enumeration of all the elements; guess, from what Holy Scripture indicates, all that is passed over in silence."[25]

Combining natural philosophy with theology became a characteristic of Christian Genesis commentaries, many of which were presented as hexaemerons. Robert Grosseteste's great *Hexaëmeron* was an encyclopedic investigation of Genesis that made use of many interpretive techniques, including literalism, Platonic formalism, and allegory. Grosseteste refers frequently to St. Basil, with whom he generally agrees.[26] For example, he relies on St. Basil for a description of Moses, noting that he was trained by Egyptian philosophers.[27]

Nevertheless, in spite of his respect for St. Basil, Grosseteste was at root an eclectic who presented ideas from a great many sources. He cited Strabus, stating, "The heaven is not the visible firmament."[28] He also was much taken by the Hebrew of the original, as explained by Jerome. He prefers Jerome's interpretation, based on his knowledge of Hebrew, to that of Basil:

> Basil explains [the spirit which moved over the waters] as follows: "He brooded over and gave life to the nature of the waters, in the way that a broody hen initiates the power of

---

24 Ibid.

25 Ibid.

26 Robert Grosseteste was a luminary of thirteenth-century thought. R. W. Southern, *Robert Grosseteste: The Growth of an English Mind in Medieval Europe* (Oxford: Clarendon Press, 1986), discusses his life and thought as does James McEvoy, *The Philosophy of Robert Grosseteste* (Oxford: Clarendon Press, 1982). C. F. S. Martin, the translator of Grosseteste's *Hexaëmeron*, sees the work as a product of his "early maturity." If so, it presents a keen and widely read mind, able to discuss, synthesize, and marshal diverse views into the service of scriptural understanding as viewed in the thirteenth century. My discussion only briefly and imperfectly touches on certain aspects of a long and complicated work. Martin's translation, used herein, is Robert Grosseteste, *On the Six Days of Creation* (Oxford: British Academy by Oxford University Press, 1996).

27 Grosseteste, *Six Days*, pt. 1, chap. 6.

28 Ibid., pt. 1, chap. 16.

> life in what she is brooding over". But Jerome says "in Hebrew the word is *merepheth*, which is what we would call brooding over or hatching, like a bird that gives lifes [*sic*] to its eggs by warming them. So we understand it[.]" Jerome goes on to say "to be said not of the spirit of the world" (that is, of the wind) "as some [Onkelos?] think, but of the Holy Spirit". Just as the will of the craftsman moves over the things that he is to make, so the Holy Spirit had it in his power how each and every thing was to be set out.[29]

Grosseteste passed on several ideas about nature and theology based on these venerable traditions.[30] He discussed the idea of how God's moving spirit worked upon the formless first matter, since "'Earth' and 'water' and 'the deep' also mean formless first matter."[31] This Platonic view did not make irrelevant the four elements and the observable natural processes of the physical world, but such things were not essential to Grosseteste's theological-philosophical point. He was open to different notions about the composition of the heavens and about the number of celestial spheres. The firmament and that above the "moon right up to above the fixed stars where the upper waters are placed" may or may not be made of a quintessence. He was quite willing to acknowledge that he did not know whether any of the philosophers had "found out the truth." Indeed, the philosophers wrote things "that are incompatible. For some of them say that there are only four elements, and that the whole which we call the firmament is of the nature of fire," while "others, like Aristotle and his followers, argue for a fifth kind of body, over and above the four elementary natures." Further discussion would "bore the listeners, and it does not seem particularly necessary. . . ."[32]

---

29 Ibid., pt. 1, chap. 20.

30 The "brooding bird" as the image of the spirit of God came to be a favorite of alchemical-chemical authors, see, e.g., Micheel Sandivogius [Michael Sendivogius], *New Light of Alchymy*, trans. J[ohn] F[rench], M.D. (London, 1674), 98–99 and Peter Severinus, *Idea medicinae philosophicae* (Basel, 1571), 42.

31 Grosseteste, *Six Days*, pt. 2, ch. 6.

32 Ibid., pt. 3, ch. vi.

III

Rabbinic thought influenced Christians in the ancient world and continued to do so, but to a lesser degree, throughout the Middle Ages. It again became important to the Christian interpretation of Genesis in the sixteenth century. Although Hebrew and the Hebrew Bible were known to and used by Christians almost from the beginning of Christian exegesis, their importance varied.[33] Nicholas de Lyra (ca. 1270–1349) used rabbinic interpreters, especially Rashi (1040–1105), in his *Postilla,* a work that influenced Luther and others.[34] This traditional interest in Hebrew and rabbinic interpretations intensified in the sixteenth century when Giovanni Pico della Mirandola discovered the Hebrew Kabbalah and Hebraic humanists, such as Johannes Reuchlin, Sebastian Muenster, Wolfgang Capito, Konrad Pellikan, and Agostino Steuco, undertook to understand the Jewish biblical tradition. The great reformers Martin Luther, Jean Calvin, and Huldrych Zwingli were also "Hebraists." The Hebrew tradition, which was believed to descend from Sinai, was seen as a pure tradition corrupted by later Jewish blindness but nonetheless valuable to a correct understanding of scripture. The belief in the Hebrew *veritas* developed by theologians made use of the traditional knowledge of nature and also influenced early modern natural philosophers in their reading of the Book of Nature.

The rabbinic tradition was built upon principles of interpretation that extended from the time of the second temple. The earliest compilation of

---

33 Beryl Smalley has detailed Christian interest in Hebrew in the Middle Ages. See Smalley, *The Study of the Bible in the Middle Ages* (Notre Dame, IN: University of Notre Dame Press, 1964).

34 De Lyra made use of natural philosophical doctrines such the elements and the crystalline spheres of the heavens in his discussion of Genesis 1. His integration of the Jewish understanding of the structure of the Hebrew scriptures was, in distinction to his use of scientific doctrines, a more original contribution to Christian exegesis. Of use in the present study was Nicholas de Lyra, *Postilla super totam Bibliam* (Nuremberg, 1481). Paul of Burgos's (1350–1435) commentary, included in that edition of the *Postilla*, discusses the gathering of the waters in Genesis, demonstrating the centrality of the biblical creation account to aspects of theoretical astronomy. See Edward Grant, *In Defense of the Earth's Centrality and Immobility*. Transactions of the American Philosophical Society, vol. 74, pt. 4 (Philadelphia: American Phiosophical Society, 1984), 23–24.

commentaries, the *Midrash Rabbah, Genesis,*[35] dates from the fifth century CE. It is largely homiletic in nature, filling in perceived scriptural blanks, expanding biblical stories, or asking questions of Genesis. It uses alphabetic interpretations, folk stories, quotations from scripture, and some pagan ideas and doctrines of nature. The homiletic use of ideas about matter is seen in *Midrash Rabbah, Genesis* 4:7, where the discussion moved from the elements to the judgments of God.

> AND GOD CALLED THE FIRMAMENT HEAVEN—SHAMAYIM (1,8). Rab said: [*Shamayim* is a compound of] *esh* (fire) and *mayim* (water).... The Holy One, blessed be He, took fire and water and beat them up together, and from them the heaven was made.
>
> SHAMAYIM is written, for they [the heavens] weigh up (*shamim*) the deeds of men: if one is worthy, *The heavens declare his righteousness* (Ps. XCVII, 6); but if not, *The heavens shall reveal his iniquity* (Job XX, 27).[36]

Perhaps the heavens are made of water:

> SHAMAYIM: [It is so called] because men wonder (*mishtommemim*) at it, saying: 'Of what (*shelmah*) is it composed? Of fire? Of water? 'Tis a mystery!'... [Scripture] comes and explains it: '*Who layest the beams of Thine upper chambers in the waters*' (Ps. CIV, 3): this shews that it is water.[37]

Or perhaps the heavens are related to fragrant spices or chemicals.

> [SHAMAYIM means like] *sammim* (chemicals): just as some chemicals are blue, others red, some black and others white, so is heaven: sometimes it is blue, sometimes red, sometimes black and sometimes white.[38]

And the rabbis compared everyday processes to the creation:

---

35 Also called *Genesis Rabbah.*

36 *Midrash Rabbah, Genesis,* trans. H. Freedman (London: Soncino Press, 1983), 1:32.

37 Ibid.

38 Ibid.

> R. Isaac said: [SHAMAYIM means] *sa-mayim*, i.e. be laden with water. Compare this to milk in a bowl: before a drop of rennet falls into it, it quivers, but as soon as a drop of rennet falls into it, it immediately curdles and stands still. Similarly, *The pillars of heaven quiver* (Job XXVI, 11); then the solidifying substance was infused into them, whereupon, '*And there was evening and there was morning, a second day*' (Gen. I, 8). That agrees with Rab's dictum: [God's] handiwork was liquid and on the second day it congealed.[39]

Rashi (R. Solomon ben Isaac [1040–1105]) relied on rabbinical tradition in composing his influential biblical commentary. After noting that Genesis is not phrased in such a way as "to teach the order of creation," Rashi stated that "water preceded the earth" and that the heavens were "created from fire and water." The purpose of Rashi's commentary was to exalt God and foster an ethical-spiritual understanding of the text; he had little or no interest in the physical processes of creation or matter theory.

Rashi's reliance on Midrash was well known to the other outstanding rabbinical commentators on Genesis. Maimonides (1134–1204) in his *Guide of the Perplexed*, a work known to most sixteenth-century Christian Hebraists, referred to the sages of *Genesis Rabbah*, who had taught that God created the luminaries on the first day, but did not suspend them in the heavens until the fourth day.[40] He then gave an interpretation of the biblical text based upon natural philosophy and the doctrine of the four elements:

> Among the things you ought to know is that the four elements are the first to be mentioned after the heaven. For, as we have said, the term *earth* mentioned in the first place applies to them. For he mentions *earth, water, spirit,* and *darkness.* Now *darkness* is the elemental fire; do not think anything else.... The elemental fire was designated by this term, because it is not luminous, but only transparent.... The elements are mentioned according to their natural position; namely, first the earth, then the water that is above

---

39 Ibid., 34.

40 Moses Maimonides, *The Guide of the Perplexed*, 2.30, trans. Shlomo Pines (Chicago: University of Chicago Press, 1963), 350.

> it, then the air that adheres to the water, then the fire that is above the air.[41]

Maimonides as a "scientific" rationalist stood apart from his younger contemporary, Ramban (1195–1270),[42] a Kabbalist with broad secular knowledge. Ramban taught that "the doings of creation is a deep secret, not to be understood from the writings [the written texts] and is not clearly understood except in light of the kabbalah (the oral mystical tradition). . . ."[43] Although Ramban believed that in essence the Torah as a whole is one name of God, he could also interpret it at a plainer, more straightforward level. Ramban understood that Rashi had limited his interpretation to the plain meaning of the text as annotated by the rabbis. Ramban, in contrast, as a master of the mystical Kabbalistic tradition, could give his "correct," deeper interpretation of the text. The "Holy One, blessed be He" created everything from nothing. This is what the verb *created* means. Yet God did not immediately create each thing from nothing; rather, He created a first matter from nothing. This first matter is what the "Greeks call hyle." From "hyle" all was formed. Ramban joined this Greek concept of the hyle to the Kabbalistic-mystical ideas of *Sepher Yetzira* and *Sepher Bahir*, which teach that the universe was made from nothing by the wonder-working word of God. God created space, and then he created the four elements. The four elements combined to fill space with visible matter, the Earth. Air, Fire, and Water also surround the Earth.

A great deal could be said about the blending of natural theory with rabbinic thought, not only by Maimonides and Ramban, but also by their fifteenth–sixteenth century successors Abravanel and Sforno.[44] A discussion of this, however, would go beyond the present purpose of introducing the Hebrew tradition of textual commentary that influenced Christian biblical commentators of the sixteenth century.

---

41 Ibid., 351.

42 Rabbi Moses ben Nahman (Nachmanides).

43 The following translations, unless otherwise indicated, are mine. The Hebrew text is from *Mikraoth Gedoloth* (Jerusalem, Isr.: B. Bruchman, 1995), 4–14.

44 I have delivered a paper on Abravanel and Sforno, but not published it:"Abravanel and Sforno, Jewish Biblical Exegesis in the Renaissance," Renaissance Society of America, Conference, March 1992.

IV

A detailed analysis of sixteenth-century Christian commentaries on Genesis is also beyond the scope of this essay.[45] A discussion of the thought of three important commentators should suffice to illustrate the fusion of the medieval tradition with humanism and with the search for the Hebrew *veritas*. The works of Giovanni Pico della Mirandola, Martin Luther, and Vatican librarian Agostino Steuco represent both the variety and continuity of sixteenth-century biblical commentary.

Grosseteste's thirteenth-century *Hexaëmeron* had been securely in the theological tradition; it used natural philosophy to illuminate Gospel truths. In the fifteenth century, Giovanni Pico della Mirandola (1463–94) employed both the medieval theological tradition and the new thought of the Renaissance. He departed from the purely theological hexaemeron in his *Heptaplus*, or commentary on the seven days of creation. (The static nature of the seventh day traditionally set it apart from the six active days of creation—in fact, it was a day without creation.) Pico, although he did not read Hebrew well, immersed himself in the rabbinic tradition.[46] He wrote his work in order to explicate the three worlds of his Christian-Platonic and rabbinic universe: the angelic, the celestial, and the sublunary. Knowledge of astronomy and of the elements being essential to comprehending the celestial and sublunary

---

45 Arnold Williams, *The Common Expositor* (Chapel Hill: University of North Carolina Press, 1948), discusses Genesis commentaries from the period 1527–1633. In the forty-plus commentaries, he found what I have also noted with regard to scientific interpretation—that science was used to support a theological purpose: "Neither in the smaller details nor in the larger underlying philosophical principles do the scientific ideas . . . excepting that of Mersenne, rise above those generally accepted by educated men of the times" (198).

46 Pico's brief scholarly life encompassed so many aspects of the apex of the Renaissance Neoplatonic revival that it is difficult to characterize. Of great importance was his study of Hebrew and, even more importantly, of Hebrew mystical texts. This has been studied in detail by Chaim Wirszubski, *Pico della Mirandola's Encounter with Jewish Mysticism* (Cambridge, MA: Harvard University Press, 1989). Pico's *Heptaplus*, his commentary on the first chapter of Genesis, was written in 1489 and reflects both his Platonism and his Hebraic learning. The English translation of *Heptaplus* by Douglas Carmichael is in *Pico Della Mirandola, On the Dignity of Man, On Being and the One, Heptaplus* (New York: Bobbs-Merrill, 1965).

realms, natural philosophy was a major focus of his commentary on Genesis. His was a philosophical, natural, theological discussion and as such bridges the older theological hexaemeric tradition and the natural, philosophical orientation of the "new science" of the sixteenth century.

Pico looked at Moses through the eyes of Plato, who was known to Renaissance humanists by Numenius's appellation, Moses Atticus.[47] The simple words of Genesis veiled from the eyes of the uninitiated the true and profound underlying "platonic" philosophy. Pico revealed the hidden knowledge in his discussion of the cultic tent or tabernacle. Its division into three parts represents the three worlds. The menorah, with its seven branches, signified the seven planets.[48] Genesis must be understood and interpreted, therefore, in terms of a trilevel cosmos. Pico did not follow the hexaemeric pattern of Basil and Augustine. He wanted a better pedagogical arrangement, which ended each discussion with the seventh day, a day focused on Christ, the Sabbath of Christians:

> Just as with Moses the seventh day is the Sabbath and a day of rest, so we have taken care that every exposition of ours shall always in the seventh chapter be turned to Christ, who is the end of the law and is our Sabbath, our rest, and our felicity.[49]

---

47 Pico's friend, Marsilio Ficino, also developed the idea that the thought of the "divine Plato" was congruent with the revelation to Moses. This is apparent in his commentary on his translation of the works of Plato, especially the *Timaeus*. There he wrote concerning the Creation, "Moyses ait, In principio fecit Deus coelum et terram. Plato ait, Deus primo ignem fecit et terram" [Moses said, In the beginning God created the heaven and the earth. Plato said, God first made fire and earth] (*Timaeum* compendium cap xxiiii, in *Omnia divini Platonis Opera*, trans. Marsilio Ficino [Venice: Iannem Mariam Bonellum, 1556], 463). Both Pico and Ficino believed that Plato was heir, in part, to God's revelation to Moses on Mount Sinai.

48 Pico della Mirandola, *Heptaplus*, second proem, trans. Carmichael.

49 Ibid. Raymond B. Waddington, "The Sun at the Center: Structure as Meaning in Pico della Mirandola's *Heptaplus*," *Journal of Medieval and Renaissance Studies* 3 (1973): 69–86, summarizes the different views of Frances Yates and Charles Trinkaus on Pico as a magus based on readings of *On the Dignity of Man* and the *Heptaplus* (Yates: Yes; Trinkaus: No). More importantly, he demonstrates that the *Heptaplus* is centered on Christ (the Sun), the creator (the Logos of the Gospel of John), and functions as a treatise on how knowledge of the levels of creation enable one to control nature: "This active

Pico's first exposition dealt with the elemental world. It focused on traditional matter theory as it applied to the sublunary world. This was the physical truth underlying Genesis. "Empty and void" referred to unshaped matter, "destitute of all form." "Darkness on the deep" implied boundless matter and privation. The earth beneath the waters represented the potential fecundity of earth when irrigated. The waters referred to the accidental qualities that attach to matter. The "Spirit of the Lord" was the efficient cause, which informed and informs the Earth. "Let the waters that are under the heaven be gathered together in one place ... and let the dry land appear," introduced matter clothed in form. Because elements can neither be seen nor touched, the passages concerned water and earth as informed or visible substances.

Pico's commentary on the elemental world is essentially platonic and devoid of explicit Christian theology until the seventh chapter. There he points to Christ as the perfect platonic form of humanity, the "perfection of all men." Man is the end and goal of creation and Jesus is the pinnacle of mankind. Pico's "*Fourth Exposition*: Of the Human World," deals with man in nature. Because man is made of materials from all three worlds, the material, celestial, and angelic, he is a microcosm of the macrocosm. Man, like the creation, consists of three parts: body, rational soul, and immortal soul.

The Divine Intellect is to man like the Spirit of the Lord that hovered over the waters. It illuminated man and made him in God's image. Pico saw spiritual and cosmic analogies in Genesis. The waters below were like the sensual-appetitive aspect of man, whereas the waters above were like the Spirit of the Lord. The sun is symbolic of the upper waters and of the way in which the lower waters can be influenced by the celestial world.[50] The elemental world and man are thus connected with the supernal spiritual world. Pico's discussion of Genesis was shaped by both his philosophical beliefs and his Christian orthodoxy. His theology was rooted in philosophy and thought about the natural world. He believed that the words of Genesis were written in a code that, when deciphered, led to insights about God, man, and the

---

operator and controller of nature is a microcosm of the creator. A magus:" (85). I agree that Pico wants to convey an understanding of nature and scripture that empowers those who understand, those who, during the Renaissance, are called magi.

50 Pico della Mirandola, *Heptaplus*, "*Fourth Exposition*," chap. 4.

structure of the cosmos. These insights agreed substantially with ancient authorities, both pagan and Christian.

Martin Luther (1483–1546) began his *Lectures on Genesis* in June of 1535. He completed the first two chapters within a few weeks. The *Lectures*, given late in Luther's life, evince at once his knowledge of Augustine, Hilary, and the medieval tradition, especially the "Hebraism" of Nicholas de Lyra and humanistic analysis. They also reflect his knowledge of natural philosophy.[51] Luther generally explicated the plain meaning of Genesis rather than revealing the secrets or allegories hidden within it. This approach allowed for a limited and theologically acceptable use of natural philosophy in order to establish the word of God.

Luther asserted that the world was 6,000 years old. The concept of an eternal world advocated by Aristotle was a product of human reason and thus, inferior to the revealed truth of Genesis. Plato, he said, may have picked up sparks of Hebraic wisdom, leading him to declare that the world had been created.[52] Plato's use of Moses was not of particular interest to Luther. He referred those seeking more detailed knowledge of the matter to the works of De Lyra. For himself, Luther generally chose the literalism of Basil over the allegory of Augustine as a key to Genesis. In all cases, however, true knowledge comes from the Holy Spirit:

> Therefore so far as this opinion of Augustine is concerned, we assert that Moses spoke in the literal sense, not allegorically or figuratively, i.e., that the world, with all its creatures, was created within six days, as the words read. If we do not comprehend the reason for this, let us remain pupils and leave the job of teacher to the Holy Spirit.[53]

As a humanistically trained textual scholar, Luther made use of Hebrew to understand the text.[54] The key terms תהו (*tohu*) and בהו (*bohu*) mean

---

51 Martin Luther, *Lectures on Genesis* trans. George V. Schick in vol. 1. of *Luther's Works*, ed. Jaroslav Pelikan (St. Louis, MO: Concordia, 1958), ix–xi.

52 Ibid., 4.

53 Ibid., 5.

54 Luther's commentary, like most humanist works, stresses language and grammar. Scholastics tended to see humanists as superficial grammarians. See Charles G. Nauert Jr., "The Clash of Humanists and Scholastics: An Approach to Pre-Reformation Controversies," *Sixteenth Century Journal*, 4 (1973): 5–18.

empty and void. This is illustrated by comparing Genesis with Isaiah 34:11, where the same words are used. The conjunction "and" used with the plural noun ואלהים (Elohim) indicated the Godhead. Luther attacks the "sophistry" of the Jews in claiming that the plural does not refer to the Trinity but is a plural of majesty:

> Therefore the feeble sophistry of the Jews is unacceptable, that the plural number is made use of for the sake of respect. What room is there here for respect? Especially since our German custom to use the plural number out of respect when we speak of only one person is not common to all languages.[55]

Luther was very traditional in exalting Moses and rejecting "the Jews." For the most part, he avoided rabbinic commentary while accepting knowledge of Hebrew as essential to understanding scripture. Luther, however, did use some ideas from Rashi that are found in De Lyra:

> Lyra approves of the opinion of Rabbi Solomon, who says that in the account of the third day this statement appears twice: "And God saw that they were very good." Therefore one of the two statements refers to the work of the second day, which was completed on the third, when the waters under the heaven were separated in a more suitable way.[56]

In keeping with his belief that revelation always trumps philosophy, Luther placed scripture above science. Whatever natural philosophy taught was inferior to divine wisdom. Indeed, Moses had not been teaching the mechanics of nature; he had been telling the story of God's creation for the faithful:

> Moreover, in the term "heaven" is included all that the philosophers divide into eight spheres, fire, and air. For on the third day he makes mention of flowing waters for the first time. It is plain that in Holy Scripture the air in which we live is called "heaven" because Scripture speaks of "the birds of the heaven." Likewise, it says that the heaven is shut

---

55 Luther, *Lectures on Genesis*, 12.

56 Ibid., 32.

> up when it is not raining; likewise, it says that the heaven rains. All this happens in the air, not in the spheres of the moon or of the other planets. Therefore, this division of the spheres is not the teaching of Moses or of Holy Scripture; but it was thought out by learned men for the purpose of teaching, something which we ought to recognize as being of great benefit.[57]

Christians, according to Luther, should recognize the usefulness of science but remember that the words of God take precedence over the wisdom and reason of men:

> We Christians must, therefore, be different from philosophers in the way we think about the causes of these things. And if some are beyond our comprehension (like those before us concerning the waters above the heavens), we must believe them and admit our lack of knowledge rather than either wickedly deny them or presumptuously interpret them in conformity with our understanding. We must pay attention to the expression of Holy Scripture and abide by the words of the Holy Spirit . . .[58]

Luther's focused Protestant study of the literal meaning of scripture stands in marked contrast to that of his Catholic contemporary Agostino Steuco.[59] Steuco, a product of Italian humanism, had emphasized the study of Hebrew. He was concerned with establishing the correct text by revealing errors through philology. He brought to his task not only the tools of humanism but also vast learning in the Christian and pagan traditions. He was a well-trained successor to Pico and Ficino. His analysis takes into account many factors that Luther considered to be, at best, extraneous to establishing the meaning of God's word.

Steuco's *Recognitio Veteris Testamenti ad Hebraicam Veritatem* and *Cosmopaeia* are detailed studies that, at times, exceed the bounds of orthodoxy. This is especially the case, according to Ronald Delph, in Steuco's

---

57 Ibid., 29.

58 Ibid., 30.

59 Ronald Delph, "Agostino Steuco and the Use of the *Hebraica Veritas* in the Counter-Reformation," unpublished paper presented at The Renaissance Society of America Conference, March 1992.

use of rabbinic sources.[60] The focus of Steuco's scholarship was the text. By comparing Jerome, the Septuagint, the Hebrew Bible, the Church Fathers, and philosophers such as Plato, Aristotle, and Hermes Trismegistus (who knew the ancient wisdom of Moses), he hoped to correctly interpret the Bible.

In the *Recognitio*, Steuco displays his vast erudition. He notes that although Aristotle did not view the world as God-created, Mercurius [Hermes] and Plato did.[61] He also discusses Origen's teaching that the world was created by Wisdom, an appellation of the Son. Origen's understanding of Wisdom as a divine manifestation agrees with the Jewish Aramaic translations of the Pentateuch. Steuco further analyzed the plural name of God, Elohim. It derives from Eloha (אלֹה) and is always used with a singular verb. This grammatical fact frustrated Christians who wanted to assert that the Trinity is unambiguously present in the Old Testament.[62] On a naturalistic note, he identified the Chaos of the Hebrews as conterminous with the prime matter of Greek philosophy.[63] Steuco turned to Basil to explain the meaning of the "Spirit of God moved." Spirit in Hebrew is *ruach* (רוח), which can mean wind or the Holy Spirit. Basil in *Contra Eunomium* taught that the Divine Spirit dispersed the air to the various parts of the creation. Following Basil's example and using the Aramaic Targum, Steuco shows that the Holy Spirit divided the water from the earth. He further notes that *Midrash Rabbah, Genesis* speaks of the Spirit as that of the Messiah.[64]

In his *Cosmopaeia*, Steuco harmonizes Anaxagoras, Hesiod, the *Timaeus*, Lactantius, Aristotle, and Cicero with Moses's teaching that "the earth was without form and void."[65] He also compares the function of chaos in Genesis to that of the heart and brain, which are superior to the other organs, as per Hippocrates. The humanist belief in the unity of all true ancient knowledge is readily apparent in Steuco's syncretistic attempt to plumb the depths of scripture. Luther's use of the plain meaning guided by the Holy Spirit is not readily apparent in the *Cosmopaeia*. The Bible text, to Steuco,

---

60 Ibid., 10.

61 Agostino Steuco, *Recognitio Veteris Testamenti ad Hebraicam Veritatem* (Venice, 1529), fol. 87r. I thank Ronald Delph for the citations and for copies of Steuco's work.

62 Ibid., 87v.

63 Ibid.

64 Ibid., 89r.

65 Agostino Steuco, *Cosmopaeia* (Venice, 1535), 10v.

could not be understood without applying the broadest possible learning. That learning included pagan literature and science.

In spite of the different approaches of Luther and Steuco, both looked to natural philosophical doctrines in their scriptural study. Like their contemporary commentators and those reaching back to Philo, they acknowledged that natural philosophy had some role to play in the theological life of Christians. The learned world had no reason to reject philosophy unless it conflicted with theological truth, as in the traditional Christian rejection of Aristotle's uncreated, eternal world. Philosophy, however, was subservient to theology even for those, such as Philo, Origen, and Steuco, who were not literalists. In fact, in his commentary on Genesis, Jean Calvin attacked Steuco, by name, as a believer in the pagan idea of preexistent matter.[66] Theology controlled how and to what extent natural philosophy was used by learned Christian scholars. Nonetheless, there was another tradition, descending from the ancient world, in which theology and scripture were secondary to the interpretation and study of nature. That tradition was alchemy, the subject of the next chapter.

---

66 John Calvin, *Commentary on Genesis*, trans. John King (Edinburgh: Calvin Translation Society, 1847), I: Gen. 1:1. For a detailed discussion of Calvin's approach to Genesis, see David C. Steinmetz, "Calvin As an Interpreter of Genesis," in *Calvinus Sincerioris Religionis Vindex*, ed. William H. Neuser and Brian G. Armstrong, Sixteenth Century Essays and Studies, vol. 36 (Kirksville, MO: Sixteenth Century Journal Publishers, 1997), 53–66.

2. Creation and the Alchemical Tradition

The development of alchemy in the west differed from that of Christian theology. Christian theology was something new, occasioned by the rise of new religion. Although Christianity made use of Jewish and pagan materials, it did so in a transformative manner. Its central creation text was Genesis and it interpreted that text according to its faith tradition. Origen, Basil, Augustine, et al. focused primarily on the spiritual meaning of creation. Pagan science, ideas of matter, and astronomy were integrated to illustrate theological principals. God's revelation always trumped reason, as the continued rejection by theologians of Aristotle's eternal cosmos demonstrates.

In contrast, what became alchemy was a complex convergence of Hellenistic and Roman creation stories, ideas about religion, matter, and how to obtain illumination and wealth by the transmutation of physical matter. Developing over the centuries, western alchemy partook of Neoplatonic, Neo-Pythagorean, Gnostic, and Jewish Kabbalistic streams of thought. These ideological traditions produced several creation texts. To be acceptable in a Christian world, those creation texts, and the alchemical ideas arising from them, had to be shown to be congruent with Genesis. Nonetheless, alchemy retained its gnostic essence, its doctrines, and its laboratory practices. The Christian Creator and the Genesis text were superimposed on alchemy without significantly altering its doctrines and practice. Indeed, alchemy's non-Christian essence helps account for its ambiguous and often heretical status in the Christian world. In the following pages I will discuss the roots and nature of alchemy and the Christianization of its views of creation.

I

The "empirical" basis of alchemy was a chemical craft tradition of making things such as glazes, metal alloys, and artificial gemstones. This tradition is found in cuneiform tablets from seventh-century BCE Assyria,[1] but can be seen in artifacts from earlier times in Egypt and the near east. Alchemists, being craftsmen, tried to obscure their trade secrets and procedures from potential competitors. They also developed routines, or rituals, designed to ensure success in their craft. This "practical" aspect of alchemy we find interpreted according to theories rooted in various creation texts and creation statements. This is not surprising given that the alchemists were working closely with the products of creation and manipulating them in ways unknown to farmers and herdsmen.

The creation stories and views of nature current in the first centuries of the Common Era were the materials from which alchemical theory was constructed. Hellenistic-Roman religion and philosophy consisted of competing and syncretistic sects identifiable as Neo-Pythagorean, Neoplatonic, Egyptian or Near Eastern, Judaic, Gnostic, and Christian, or combinations of the foregoing. From our perspective, these intellectual spiritual pathways seem to have much in common. Their adherents, however, were keenly aware of their differences. They often defended the correctness of their own faith and strenuously opposed others' as corrupt and false visions of life and salvation.

Neoplatonists and Neo-Pythagoreans shared a view of the cosmos as a created, mathematical harmony. For them, Plato's *Timaeus* was a central text. It argued that God created the material world according to a perfect, unchanging, formal pattern. The world was created to be good. Yet perfect form translated into matter produced not eternal material beings but imperfect creatures subject to change. While the material world remained one of elements and change, the soul had a different, eternal nature. It could return to the perfect realm, the pattern of creation.

Concerning the elements, Plato has Timaeus say:

---

1 R. Campbell Thompson, "A Survey of the Chemistry of Assyria in the Seventh Century B.C.," *Ambix* 2 (1938): 3–16.

> In the first place, we see that what we just now called water, by condensation, I suppose, becomes stone and earth, and this same element, when melted and dispersed, passes into vapor and air. Air, again, when inflamed, becomes fire, and, again, fire, when condensed and extinguished, passes once more into the form of air, and once more, air, when collected and condensed, produces cloud and mist—and from these, when still more compressed, comes flowing water, and from water comes earth and stones once more—and thus generation appears to be transmitted from one to the other in a circle. Thus, then, as the several elements never present themselves in the same form, how can anyone have the assurance to assert positively that any of them, whatever it may be, is one thing rather than another? No one can. . . .[2]

Plato's elements, which constantly change form, became a part of the alchemical understanding of nature. The creation or formation of the material world was viewed in the processes of condensation and rarefaction, so noticeable in the artisans' use of fire.

By the first centuries of the Common Era, when alchemy proper arose, the philosophical schools of Neoplatonism and Neo-Pythagoreanism had adopted religious practices and ideas and become religious movements.[3] Although maintaining the foundation of classical Platonic and Pythagorean texts, they also looked to pagan and pagan-Jewish oracular texts for spiritual-mystical insights. For example, the *Rhapsodies*, also known as the *Rhapsodic Theogony* or *Sacred Discourse in Twenty-Four Rhapsodies*, attributed to Orpheus, and the *Chaldaean Oracles* were sacred texts used by Plotinus, Iamblicus, Porphyry, and Proclus.

The *Rhapsodies* presents a creation that incorporates accounts of the classical gods and their forebears. Time generated the *aither* of the atmosphere

---

2 Plato *Timaeus* 49, trans. Benjamin Jowett in *The Collected Dialogues of Plato*, ed. Edith Hamilton and Huntington Cairns (New York: Pantheon Books, 1961), included in *The Alchemy Reader: From Hermes Trismegistus to Isaac Newton*, ed. Stanton J. Linden (Cambridge: Cambridge University Press, 2003), 32–33.

3 Neoplatonism is the philosophical-religious system developed by Platonists while Neo-Pythagoreanism developed Pythagorean numerical and harmonical ideas. These philosophies and their religious manifestations are discussed in Brian P. Copenhaver, *Hermetica* (Cambridge: Cambridge University Press, 1992), xiii–xxxii.

and the boundless or chasm. Time formed an egg from *aither*; the egg produced Phanes or Eros. Phanes mated with his daughter, Night, who gave birth to Uranos (Sky) and Ge (Earth). Uranos and Ge gave birth to the three Fates and the Titans. The story of Kronos and Zeus follows in due course. Within *The Rhapsodies* and other Orphic works, the Neoplatonic/Neo-Pythagorean religion discovered and developed profound allegorical truths that fit with the writings of Plato and the Pythagorean tradition.

The *Chaldaean Oracles*, traditionally attributed to Zoroaster, were thought to contain the wisdom tradition of the near eastern magi, part of the tradition found in the *Timaeus*. Indeed, Proclus is quoted as having said: "If I had in my power, out of all the ancient books I would suffer to be current only the *Oracles* and the *Timaeus*."[4] The *Oracles* were said to be the redaction of Zoroastrian oracles made by the second-century "Chaldean" wise man Julian. Inherent in the belief in Chaldean wisdom was the idea that such knowledge could lead to a manipulation of nature or theurgy (a term used often in the Renaissance). In his *The Mysteries of Egypt*, Iamblicus wrote of magical ceremonies as "the religious fulfillment of ineffable acts whose results outdo any effort of intellect"(2.11). The efficacy of magical acts became a staple of alchemy.

The creation account in the *Chaldaean Oracles* describes the One giving rise to multiplicity and ultimately to the material world from the "matrix." This process was poetically rendered in Thomas Stanley's seventeenth-century translation of the *Oracles*:

> MONAD, DYAD. TRIAD.
> Where the Paternal Monad is.
> The Monad is enlarged, which generates TWO.
> For the Dyad sits by him, and glitters with Intellectual Sections.
> And to Govern all things, and to Order all things not Ordered,
> For in the whole World shineth the Triad, over which the
> Monad Rules.[5]

---

4 Marinus, *Life of Proclus*, trans. Kenneth Sylvan Guthrie, ed. David K. Fideler (Grand Rapids, MI: Phanes Press, 1986), 55.

5 *The Chaldaean Oracles as Set Down by Julianus, translated into the Latin by Francesco Patrizzi and Translated into English by Thomas Stanley* (Berkeley Heights, NJ: Heptangle Books, 1989), 3; *The Chaldaick Oracles of Zoroaster and his Followers . . .* , ed. and trans. Thomas Stanley (London, 1661), http://www.esotericarchives.com/oracle/ oraclesj.htm.

The Three ruled by One preceded the differentiation of spiritual things that were to become physical:

> This Order is the beginning of all Section.
> For the Mind of the Father said, that all things be cut into three,
> Whose Will assented, and all things were divided.
> For the Mind of the Æternal Father said into three,
> Governing all things by Mind.
> And there appeared in it (the Triad) Virtue and Wisdom.
> And Multiscient Verity.
> This Way floweth the shape of the Triad, being præ-existent.
> Not the first (Essence) but where they are measured.
> For thou must conceive that all things serve these three Principles.[6]

The sacred principles operate on the matrix (undifferentiated matter) with fire and make the world:

> The first course is sacred, but in the middle.
> Another the third, ærial; which cherished the Earth in fire.
> And Fountain of Fountains, and of all Fountains.
> The Matrix containing all things.
> Thence abundantly springs forth the Generation of multivarious Matter.
> Thence extracted a prester the flower of glowing Fire,
> Flashing into the Cavities of the World: for all things from thence
> Begin to extend downwards their admirable Beams.[7]

This creation by emanation and fire was a mystical, geometrical, chemical process. The creation in the *Chaldaean Oracles* differed from creation *by* the Logos, the Creator-god of the *Sibylline Oracles.* Creation by the Logos was a concept popular among pagans and Christians. The sibyls were ancient pagan prophetesses, but parts of this work probably originated as heterodox Jewish writings, which were later edited and amended by Christians in the fifth century CE. The *Sibylline Oracles* assumed the biblical

6 Ibid., 34.

7 Ibid., 4.

creation, but imbedded it in a pagan philosophical framework that universalized its ideas beyond Judaism. The Sibyl told "truly how into being came the world" by the "Word [Logos] of God, the King." In what scholars believe to be among the later additions to the *Oracles*, she describes:

> First now God urges on me to relate
> Truly how into being came the world.
> And thou, shrewd mortal, prudently make known,
> Lest ever thou should'st my commands neglect,
> The King most high, who brought into existence
> The whole world, saying "Let there be," and there was.[8]

The Sibyl then recounted how the Logos constructed a garden of life around Tartarus, Hades:

> For he the earth established, placing it
> Round about Tartarus, and he himself
> Gave the sweet light; he raised the heaven on high,
> Spread out the gleaming sea, and crowned the sky
> With an abundance of bright-shining stars,
> And decked the earth with plants. . . .
> These by his word he made, and every thing.[9]

This Creation "Oracle," in which God speaks, and by his word makes all, paraphrases Genesis and is similar to Proverbs 3:19, which states, "[T]he Lord by wisdom hath founded the earth." Further, the Oracle's Logos could easily be viewed as a pagan analogue of the Christian creation text in the Gospel of John. John's Logos was with God and was a god. As the primordial creative manifestation of the Father, the Word was with God at the Creation. The rabbis, in a like manner, taught that the Torah was with God at the Creation. The Torah was the pattern of the created world. The Sibylline creation was consistent with the Judeo-Christian tradition that came to dominate the late Roman Empire. Thus, many of the *Sibylline Oracles* were preserved and their mixture of pagan and Judeo-Christian ideas remained a source for alchemical thought.

---

8 *The Sibylline Oracles*, bk. 1, trans. Milton S. Terry, new edition revised after the text of Ruch (New York: Eaton and Mains, [1899]), 9, http://www.sacred-texts.com/cla/sib/sib.pdf.

9 Ibid.

The complexity of the religious-philosophical schools, practices, and points of view in the early centuries of the Common Era was also manifested in a number of Gnostic-hermetic writings and in the development of the Jewish esoteric tradition, the Kabbalah. These, in turn, became sources of alchemical theory. The *Corpus Hermeticum* was the great repository of the Gnostic-alchemical texts of late antiquity.[10] It is peppered with creation language. The eleventh treatise, "Mind to Hermes," contains the words "God makes eternity; eternity made the cosmos;" and "eternity, therefore, is in God, the cosmos in eternity."[11] From late antiquity through the Renaissance, many people found a consistency between hermetic statements, the *Chaldaean Oracles*, and Plato. Hermetic documents, therefore, were considered a valuable, although imperfect, pagan transmission of ancient revealed truth. The *Corpus Hermeticum* taught that there is a knowledge or gnosis hidden in its unusual terminology and metaphors. Pagans, Jews, and Christians could all seek such gnosis by reading their holy texts and rituals in light of hermetic truths. Thus, there could be, and were, pagan, Jewish, and Christian Gnosticisms.

The creation stories in the *Corpus Hermeticum*'s treatises *Poimandres* and *Asclepius* resonated with many religious seekers. *Poimandres* presents a pagan-Gnostic creation story that reflects an awareness of Genesis and Christianity. Poimandres, or the Mind of Sovereignty, reveals that

> I am the light you saw, mind, your god . . . who existed before the watery nature that appeared out of the darkness. The lightgiving word who comes from mind is the son of god.[12]

Poimandres also describes the creation of the elements; a creation of declining perfections, from the perfect God, through the Mind, and by the seven governors of the sensible world:

---

10 Its translation by Ficino (1462–63) and publication in 1471 was a major moment in western thought because it helped to bring together Neoplatonism, Neo-Pythagoreanism, Kabbalah, and Gnosticism into a single intellectual fusion.

11 Copenhaver, *Hermetica*, 37. Copenhaver provides a detailed and eloquent introduction to the milieu of the *Corpus Hermeticum*. He discusses politics, pagan religion, religious texts, and the responses to them by several Christian fathers.

12 Copenhaver, *Hermetica*, 2.

> "The elements of nature—whence have they arisen?" I asked.
>
> And he answered: "From the counsel of god which, having taken in the word and having seen the beautiful cosmos, imitated it, having become a cosmos through its own elements and its progeny of souls. The mind who is god, being androgyne and existing as life and light, by speaking gave birth to a second mind, a craftsman, who, as god of fire and spirit, created seven governors; they encompass the sensible world in circles, and their government is called fate."[13]

The creation in *Asclepius* is very similar to that in *Poimandres.* God created all, but lesser deities, such as the heavens, administer descending realms of creation.

> [A] continuous influence carries through the world and through the soul of all kinds and all forms throughout nature. God prepared matter as a receptacle for omniform forms, but nature, imaging matter with forms by means of the four elements, causes all things to reach as far as heaven so that they will be pleasing in the sight of god.[14]

The hermetic revelations about matter, form, and creation occasioned severe reactions, not only from Christians (St. Augustine made the hermetic treatise *Asclepius* the best-known Gnostic text in the Latin West through his polemic against heresies), but also from "orthodox" pagans. Neoplatonists, who believed that Plato was divinely inspired, found inconceivable the idea that there was something beyond his intellectual-spiritual depths. Plotinus considered that gnosticism was a perversion to be opposed and defeated. Porphyry wrote in his *Life of Plotinus*:

> There were in his time many Christians and others, and sectarians who had abandoned the old philosophy, men . . . themselves deceived and deceiving many, alleging that Plato had not penetrated to the depths of intelligible reality. Plotinus hence often attacked their position in his lectures,

---

[13] Ibid.

[14] Ibid., 68.

> and wrote the treatise to which we have given the title "Against the Gnostics."[15]

The battle against esoteric knowledge continued even after the triumph of Christianity over all other religions and philosophies in the Roman Empire. The Christian Empire marked an apparent end of both paganism and Gnosticism, but traces of both would persist, especially in the alchemical tradition.

Judaism also reacted to Gnostic thought and its division of the one God into various hypostases and lesser governors. There is a story in the Babylonian Talmud that relates to secret knowledge as a source of heresies. Four rabbis entered *Pardes*, itself a Persian word. One died; one became mad; the third glimpsed the angel Metatron, mistook him for God, and became a dualistic heretic. Only one, Akiva, departed undamaged.[16] This story demonstrates a rabbinic ambivalence (Akiva did survive) regarding the kind of Gnostic thought that also upset Porphyry and Augustine.

Within Christianity, Gnosticism had its own tradition. Works such as the *Gospel of Thomas*, with its "secret words" of Jesus, were cherished by Gnostic Christians. The Nag Hammadi Gnostic texts discovered in 1947 include a rendering into Coptic of *Asclepius* on the mystical union of man and woman.[17] Eventually crushed as heretical, Gnostic sects, such as Catharism, would resurface from time to time as Christian heresies.

Judaism absorbed certain aspects of Gnosticism. These Kabbalistic Gnostic ideas were thus available to diligent Christians who were inspired to look at Jewish sources throughout the Middle Ages and the Renaissance. The rabbis taught that the Kabbalah was hidden knowledge, revealed at Sinai, and transmitted, or re-revealed, to the select few. It is impossible to securely chronicle the history of the various texts and the mystical and intellectual teachings that form the basis for pre-Zoharic Kabbalah.[18] Most scholars believe that the oldest Kabbalistic text is *Sefer Yetzira* or *Book of Formation*, attributed to the Patriarch Abraham. It has been dated to as early as the fourth

---

15 James M. Robinson, introduction to *The Nag Hammadi Library*, ed. Robinson (San Francisco: Harper and Row, 1977), 8. Robinson notes and translates Plotinus's reaction against the Gnostics and Porphyry's attack on all non-Platonists in his *Life of Plotinus*.

16 Talmud Bavli, *Haggigah* 14a–14b.

17 Robinson, *Nag Hammadi Library*, 300.

18 Pre-thirteenth century.

and to as late as the seventh century CE. It is a Neo-Pythagorean commentary on the underlying meaning of Creation. Everything is created from, or by means of, the Hebrew alphabet, in which the letters are also numbers. Each ideal number/letter connects with the perceptible, material world. The Hebrew word *sfr* in its various verb forms means "to count," "to write," "to tell." As a noun form, it may be a book, a counting, a retelling. In *Sefer Yetzira*, it becomes a series of countings or creative hypostases, the Sefirot (*ot* being the feminine plural ending):

> The Sefirot[19] made of Nothing
> The first: The Spiritwind of Living God
> The second: Wind from the Spiritwind
> The third: Water from the Wind
> The fourth: Fire from the Water
> and height and depth
> and East and West
> and North and South.[20]

Yetzira's integration of creation by emanation and of the elements of classical matter theory into the Hebrew tradition was important as a model for connecting heterodox ideas to a traditional orthodox theological tradition. This kind of justification was important to alchemy, which was the major Neoplatonic/Neo-Pythagorean Gnostic tradition to survive in a Christian world. At a later date, in the Renaissance, *Sefer Yetzira* reappeared and pollinated the reborn Platonic-Pythagorean philosophies.

II

The practical-transmutational ideas, as well as the theoretical framework, of western alchemy partook of the various spiritual and mystical traditions discussed above and placed them in the context of classical matter theory. The four elements of classical matter theory—earth, water, air, and fire—lay at the

---

19 Sefirot are the Hebrew equivalent of the demiurges or hypostases of Neoplatonism and gnosticism who connect divine perfection to the imperfect, material world

20 "Sefer Yetzira," trans. Harris Lenowitz, in *Origins: Creation Texts from the Ancient Mediterranean*, ed. and trans. Charles Doria and Harris Lenowitz (Garden City, NJ: Anchor Books, 1976), 60.

root of the concept of transmutation. By altering elemental balance, one substance could be changed into another.

The elements and the possibility of transmutation could also be seen as harmonious with atomism. Democritus (Demokritos of Abdera, d. ca. 376 BCE) developed atomism, and in the first century before the common era, Lucretius cast Democritus's philosophy into Latin verse in *De rerum natura.* The theory of atomism was well known in the ancient world and available in early modern times. Aristotle criticized it in several places and Cicero emphasized its underlying atheism in *De natura deorum*. Although it has been widely taught that the technical knowledge of atomism dates to the discovery of a manuscript of Lucretius's *De rerum natura* in 1417 and its publication in 1473, William Newman has shown that atomism was already a part of matter theory and was used by alchemists by the late thirteenth century.[21] In challenging non-atomistic theory, Lucretius questioned, "[I]f all things are created of four and all again dissolved into four, how can they be called primal gems?"[22]

Alchemists mixed ancient elemental theories with ideas of harmony, number, secret knowledge, and divine revelation. The doctrine of four elements and its criticism by atomists influenced alchemy from its earliest extant texts (third century CE) through the eighteenth century. The admixture of physical matter theory with spiritual illumination produced a tradition that could be applied to gaining wealth through transmutation as well as to providing a pathway to Gnosis and divine insight.

The earliest western alchemical texts ascribed the beginning of the art to the Egyptians, especially to the goddess Isis, although Democritus, Plato, Moses, Iamblicus, and others are also cited as masters of the art and authors of alchemical texts. One of the earliest texts is a pseudepigraph attributed to

---

21 Danton B. Sailor, "Moses and Atomism," *Journal of the History of Ideas* 25 (1964): 3–16, discusses the rise of atomism in Renaissance and early modern Europe. William R. Newman, *The Summa perfectiones of Pseudo-geber* (Leiden: E. J. Brill, 1991), demonstrates the use of atomism in the medieval alchemical tradition. The development of matter theory in light of modern research is the subject of Christoph Lüthy, John E. Murdoch, and William R. Newman, eds., *Late Medieval and Early Modern Corpuscular Matter Theories* (Leiden: E. J. Brill, 2001).

22 Lucretius, *Of the Nature of Things*, trans. William E. Leonard (London: E. P. Dutton, 1921), 29.

Democritus written before 250 CE.[23] This work, titled *Physical and Mystical Matters*, is composed largely of practical recipes but also contains an important section on the mystical master-student tradition. Within that section, the aspiring adept learned that one must harmonize the natures of the substances treated and that nature itself is an alchemical process. All this was learned through the aid of a dead master:

> Recollecting these ideas of our ordained master and knowing the diversity of matter we endeavoured to bring the natures into harmony. But as our master died before we were initiated and at a time when we had not attained the complete knowledge of matter, we were told that we must evoke him from Hades.... And as I called upon him many times, asking now how I should harmonise the natures.... And he only said "the books are in the temple". None of us knew anything about these books.... As we had completed the combination of matter, and it was time for a ceremony in the temple, we made a common feast. Then as we were in the temple one of the columns suddenly opened but we saw nothing within.... [L]eaning forward we saw with astonishment that nothing had escaped us except the precious formula which we found there:
>
> "The Nature rejoices in the nature and the nature conquers the nature and the nature masters the nature."
>
> We were much surprised that he had summed up all his writings in so few words.[24]

Although this was not a creation story, the ideas, especially that of nature rejoicing in nature, implied a unity underlying the interaction of substance and spirit. This pagan, Egypto-Greco vision of a dynamic, spiritual nature was developed through analogies in other texts. Maria the Jewess, whose works are lost, is quoted to the effect that if you "[j]oin the male and the female you will find what you seek."[25] The union of male and female is a theme of the Hermetic *Asclepius*. The dynamic force in nature, the essence of creation, is the spiritual-material union of male and female.

---

23 F. Sherwood Taylor, "The Origins of Greek Alchemy," *Ambix* 1 (1937): 37.

24 Ibid., 37–38.

25 Ibid., 42.

The creative power of nature could also be represented by the Ouroboros, the image of the snake eating its own tail. The *Chrysopoeia of Kleopatra* labels the Ouroboros "One is all."[26] Another Greek alchemical work, "The Visions of Zosimos," identifies the waters and movement, the separation of spirit from body, as the key to creation and the art. The multiplicity is really one:

> The composition of waters, the movement, growth, removal, and restitution of corporeal nature, the separation of the spirit from the body, and the fixation of the spirit on the body are not due to foreign natures, but to one single nature reacting on itself, a single species, such as the hard bodies of metals and the moist juices of plants.[27]

Alchemy assimilated various creation images and analogies, which Christian adepts believed were harmonious with the account in Genesis. The egg, for example, became a symbol of the creation, the levels of the created world, and the alchemical process. Also of significance for Christians in adopting ancient non-Genesis creation stories and matter theory was the association of Toth, or the thrice-great Hermes, with alchemy. Hermes was one of the transmitters of revealed knowledge from the Patriarchs to the pagan world. The *Corpus Hermeticum*, as the composition of a holy pagan, became not merely part of the alchemical art, but a central source for understanding the natural and spiritual worlds. When the west reacquired the *Corpus Hermeticum* at the end of the fifteenth century, its Gnostic-wisdom tradition of the first centuries of the Common Era almost immediately became linked to Renaissance philosophical endeavors. Neoplatonists saw that their philosophy had common roots with Hermeticism. They believed that their philosophy came from Moses and the Patriarchs, whose prophetic wisdom had been made known to ancient pagans.

A seminal text linking Hermes to alchemy, and known before the *Corpus Hermeticum*'s rediscovery, was the *Emerald Table*. Like *Sefer Yetzira*, its dating is difficult, but medieval and Renaissance alchemists believed it to be an authentic ancient hermetic work. It presents a "Gnostic" creation of the

---

26 Ibid., 43.

27 "Zosimos, *The Visions of Zosimos*," translation and prefatory note by F. Sherwood Taylor. *Ambix* 1 (1937): 88.

world by the "One" as both a cosmogony and as an outline of how to reproduce nature in the laboratory. It is Gnostic because in it Hermes, the adept, reveals the secrets of the creation to the initiated. From the early Middle Ages, virtually all western alchemists knew the *Emerald Table*. Its thirteen statements are a brief alchemical theology. Theologically, it is an inversion of St. Basil's *Hexaemeron*, in that natural processes are the bases of theology rather than theology the a priori arbiter of how to understand nature. In its own way, the *Emerald Table* was as influential among the adepts as the hexaemeric tradition was among theologians.

The following are the thirteen statements of the *Emerald Table*:

> The Precepts of Hermes, engraved upon the Emerald Tablet:
>
> 1. Speak not fictitious things, but that which is certain and true.
> 2. What is below is like that which is above, and what is above is like that which is below, to accomplish the miracle of one thing.
> 3. And as all things were produced by the one word of one Being, so all things were produced from this one thing by adaptation.
> 4. Its father is the sun, its mother the moon; the wind carried it in its belly, its nurse is the earth.
> 5. It is the father of perfection throughout the world.
> 6. The power is vigorous if it be changed into earth.
> 7. Separate the earth from the fire, the subtle from the gross, acting prudently and with judgement.
> 8. Ascend with the greatest sagacity from the earth to heaven, and then again, descend to the earth, and unite together the powers of things superior and things inferior. Thus you will obtain the glory of the whole world, and obscurity will fly far away from you.
> 9. This has more fortitude than fortitude itself; because it conquers every subtle thing and can penetrate every solid.
> 10. Thus the world was formed.
> 11. Hence proceed wonders, which are here established.
> 12. Therefore, I am called Hermes Trismegistus, having three parts of the philosophy of the whole world.

13. That which I had to say concerning the operation of the sun is completed.[28]

In the eyes of Christian (and later in the eyes of Islamic) alchemists, the Gnostic "unity" was actually the true God. An adept could accept the *Table* and remain religiously orthodox. Christian adepts could harmonize alchemical theory with the Trinity, Genesis, and church dogma. No matter what theology the alchemist brought to the texts, Christian or Islamic, the essential alchemical understanding of the elements of nature and its processes remained the same. The cosmic world of the One was apparent in processes on the earth and in man. The transformative powers of the sun, the moon, the wind, and other forces of nature could be mimicked in the alchemical laboratory. The adept glimpsed the creation in the alchemical processes of separation and purification. Christian adepts used the *Table* as a key to physical and spiritual knowledge. They interpreted Genesis as describing the distillations and sublimations that they believed originally formed the world and that could still be observed in the laboratory.[29]

George Ripley, a fifteenth-century English alchemist and canon of Brydlynton, illustrates the fusion of alchemical doctrines with Christian piety in his preface to the Twelve Gates of *The Compound of Alchymie*.[30] The first stanza is a Gnostic doxology addressed to the incomprehensible triune God who purifies souls:

O Hygh Yncomprehensyble and gloryous Mageste,

---

28 This translation is from John Read, *Prelude to Chemistry* (1936; Boston, MA: MIT Press, 1966), 54. E. J. Holmyard, *Alchemy* (Harmondsworth, UK: Penguin Books, 1957), 97–98, reproduces a translation by R. Steele and D. W. Singer and comments on the *Table*'s cryptic nature. The Steele-Singer translation and other texts that illustrate many of my points are found in Linden, *The Alchemy Reader*.

29 Allen G. Debus presents an extensive discussion of alchemy-chemistry and the creation in Allen G. Debus and Robert P. Multhauf, *Alchemy and Chemistry in the Seventeenth Century* (Los Angeles: William Andrews Clark Memorial Library, 1966). A more specific discussion of the importance of alchemical ideas in the literature of the sixteenth and seventeenth centuries, indeed, from Chaucer to Milton, is found in the appendix "Creation and Alchemy" in Michael Lieb, *The Dialectics of Creation* (Boston, MA: University of Massachusetts Press, 1970).

30 George Ripley, *The Compound of Alchymie* in Elias Ashmole, *Theatrum Chemicum Britannicum* (London, 1652); repr., The Sources of Science, no. 39 (New York: Johnson Reprint, 1967), 107–93.

> Whose Luminos Bemes obtundyth our speculation;
> One-hode in Substance, O Tryne hode in Deite,[31]
> Of Hierarchycall Jubylestes the gratulant gloyfycation;
> O pytewouse puryfyer of Soules and puer perpetuation;
> O deviaunt fro danger, O drawer most deboner;
> Fro thys envyos valey of vanyte, O our Exalter.[32]

The stanzas that follow constitute a prayer from a novice for divine instruction:

> O Power, O Wysdom, O Goodnes inexplycable;
> Support me, Tech me, and be my Governour,
> That never my lyvyng be to thee dysplycable,
> But that I aquyt me to thee as a trew professor:
> Att thys begynnyng good Lord here my prayer;
> Be nygh with Grace for to enforce my wyll,
> Graunt well that I may my entent fulfill.[33]

For only a pure soul could access the wisdom necessary to transform matter:

> And for soe much as we have for thy sake
> Renowncyd the World, our Wylls, and the Fleshys Lust,
> As thyne owne wylfull professyors us take;
> Syth in thee only dependyth all our trust,
> We can no ferther, to thee enclyne we must:
> Thy secret Tresorars, vouchsafe to make us,
> Show us thy Secrets, and to us be bounteous.[34]

The secrets of the art were revealed in Genesis. Ripley explains Genesis and its connection to the secret of the philosophers' stone with its power to heal and transmute:

> In the begynnyng when thou madyst all of nought,
> A globose Mater and darke under confusyon,
> By thee Begynner mervelously was wrought,
> Conteynyng naturally all thyngs withoute dyvysyon,
> Of whych thou madyst in six Dayes dere dystynction;
> As Genesys apertly doth recorde

31 One-hood in Substance, O trine-hood in Deity [cf. childhood].

32 Ibid., 121.

33 Ibid.

34 Ibid., 122.

Then Heavyn and Erth perfeytyd were wyth thy word.
So thorow thy Wyll and Power owte of one Mase
Confusyd was made all thyngs that being ys;
But yn thy glory afore as maker thou was,
Now ys and shall be wythout end I wys:
And puryfyed Sowls upp to thy blys
Shall come a pryncyple, thys may be one,
For the declarying of our *Stone*.[35]

The art of alchemy worked on the elements to imitate, through art and craft, the creative process of Genesis. By balancing the elements and the formative sexual forces, matter could be altered, thereby producing the perfectly proportioned Stone. The adept learns the secret of making the Stone when he understands the correspondence of the great world of the Creation with the lesser world of the laboratory, overseen by the true microcosm, Man. The "Stone" or end of the art is also a microcosm, but one that contains cosmic powers. Ripley believed,

In *Phylosophers* Bokes therefore who lust to se,
Our *Stone* ys callyd the *lesse World* one and three,
Magnesia also of *Sulphure* and *Mercury*,
Propotionat by Nature most perfytly.[36]

To obtain spiritual power, an adept also needed to employ the sexual forces inherent in the physical world. The *Splendor Solis* of Salomon Trismosin relies on the male-female dynamic that shaped matter.[37] The male-female forces are viewed not as superior and inferior, as they are in medicine of the period and in Aristotelian sex and social theory, but as equally potent forces, working together. Male and female interact in order to bring about change in matter. The *Splendor Solis* illustrates the interplay of male and female by showing a knight in armor, with a drawn sword, standing astride two foundations. His shield is inscribed, "Ex duabos aqui unam facite. . . ." These words, "from two waters is made one," refer to the two fountains, the right capped by a nude boy and running black fluid, and the left capped by a

35 Ibid., 122.

36 Ibid., 123.

37 Salomon Trismosin (pseudonym of an alchemist named Pfeffer), *Splendor Solis* (1582), trans. J. K. (London: Kegan Paul, Trench, Truber, and Co., 1921), also published in *Aureum Vellus order Golden Shatz* (Rorschach, 1598).

nude girl and running golden fluid. The golden (female) fluid transmuted the black (male) fluid into a golden stream that ran from the base of the fountain into a lake of gold, disappearing in the distance. The secret of transmutation is the ability to manipulate the dark male matter by means of the solar female. If all goes well, the Stone is formed. The perfect union of the sexes created the hermaphrodite, having the power to transmute matter.[38]

Alchemical sexual imagery commonly represented the Sun as a king and the Moon as a queen. The Sun and the Moon, husband and wife, denoted earth and water. They also represented the connection between the celestial and terrestrial realms. Further, they provided the model to be reproduced in the adept's furnace. Man and wife, Sun and Moon, produced four children: one boy Hot, one boy Cold, one girl Moist, one girl Dry. These children need to be altered to turn base metal into noble.

The alchemist did not seek to open new frontiers of knowledge. He looked to ancient texts for revealed truth. He did not doubt that former adepts had been successful in transmuting matter and finding the Stone, but they had written obscurely about their manipulations. Yet, the most ancient text of all, the Bible, could reveal, and be revealed, in synchronicity with alchemy. In the early seventeenth century, the alchemical theory-theological tradition was presented in an illustrated alchemical hexaemeron, *Janitor Pansophus* [The doorkeeper of all wisdom].[39] Whereas Christian theological hexaemerons interpreted nature according to scripture, the *Janitor* interpreted scripture according to alchemical tradition. The work connected the *Emerald Table* with Genesis. Figure 2 in the *Janitor* presented the elemental world and various aspects of it with a commentary based on the *Corpus Hermeticum* and the *Emerald Table* (see also *fig. II*). Figure 3 interpreted the six days of creation alchemically. The illustrations of figure 3 and the brief scriptural passages

---

38 Michael T. Walton, Robert M. Fineman, and Phyllis J. Walton, "Holy Hermaphrodites and Medical Facts: The Depiction of Hermaphrodites in Alchemy and Medicine," *Cauda Pavonis: Studies in Hermeticism*, n.s., 18 (Spring–Fall 1999): 32–36.

39 *Janitor Pansophus* in *Musaeum Hermeticum* (Frankfurt: L. Jennisii, 1625) and in *Musaeum Hermeticum Reformatum et Amplificatum* (Frankfurt: Hermannum á Sande, 1677). The plates used to illustrate *Janitor Pansophus* when it was printed were first published in *Dyas Chymica Tripartita* in 1625. They were combined with the *Janitor* text in the first edition of the *Musaeum Hermeticum*.

beneath them are connected to a Latin explanatory text taken from Ripley's preface to *The Compound of Alchemie* (see *fig. III*).

Figure 3 is a true hexaemeron. It consists of ten illustrations depicting the world as it stood on different days or at different stages of creation. The first illustration is of the Tetragrammaton, the Alpha and Omega, which, like the *Ein soph* of the Kabbalah, existed before the creation. The second is of Genesis 1:2, "the earth was without form and void and darkness was above the face of the deep." The third continues Genesis 1:2, "the Spirit of the Lord hovered over the waters." The fourth, Genesis 1:3, "God made light and it was and God saw the light that it was good." The fifth is "He divided light from Darkness and He called light day and darkness night and it was so, one day." The final five show the waters below; the dry land and plants; the Earth-centered cosmos; the animals; and, finally, man and woman.[40]

Beneath the illustrations is the text from Genesis 1:6–31. Genesis is supplemented by quotations from Psalms 33, 48, 104; Daniel 3:59–60 and The Wisdom of Ben Sira 43:1 (Vulgate version). The reader is meant to understand scriptures in light of Ripley's preface.

The *Janitor* demonstrates the stability of the hoary alchemical theory-theology in Europe. Traditional pagan or pagan-influenced texts were perceived as part of a revelation of nature, which dated to the patriarchs and Sinai. Ideas in various texts from pagan and Christian authors were believed to be harmonious because their ultimate source was God. Like the theological tradition of the church, alchemy was also a source of divine wisdom and of power to emulate God's creative acts. With the intellectual spiritual ferment of the sixteenth century, the purity of the transmission of Christian theology was questioned by the reformers. In a like manner, the new chemists, Paracelsus and his spiritual disciples, called into question the traditional formulation of alchemical doctrines. This chemical reformation and its use of Genesis to create or "restore" a pure, Christian chemical philosophy is the subject of the following chapters.

40 Translations of the captions of the illustrations are mine.

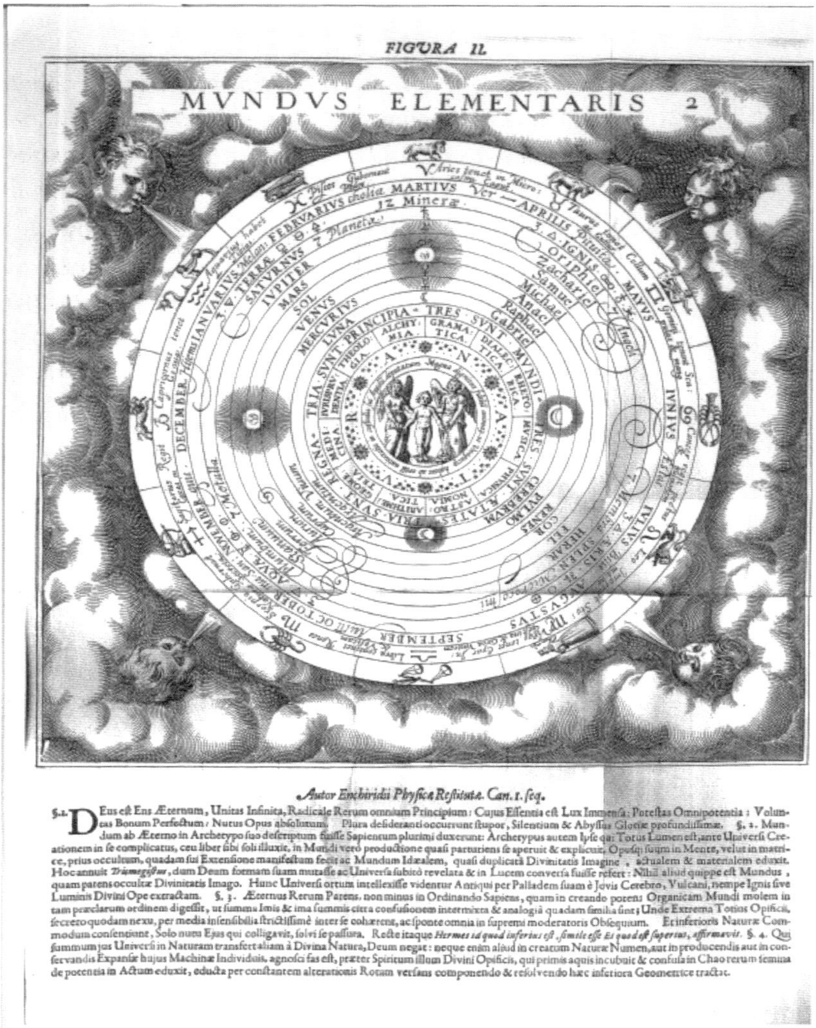

Figure II. Mundus Elementaris. *Janitor Pansophus,* figure 2 in *Musaeum Hermeticum.* Frankfurt, 1677.

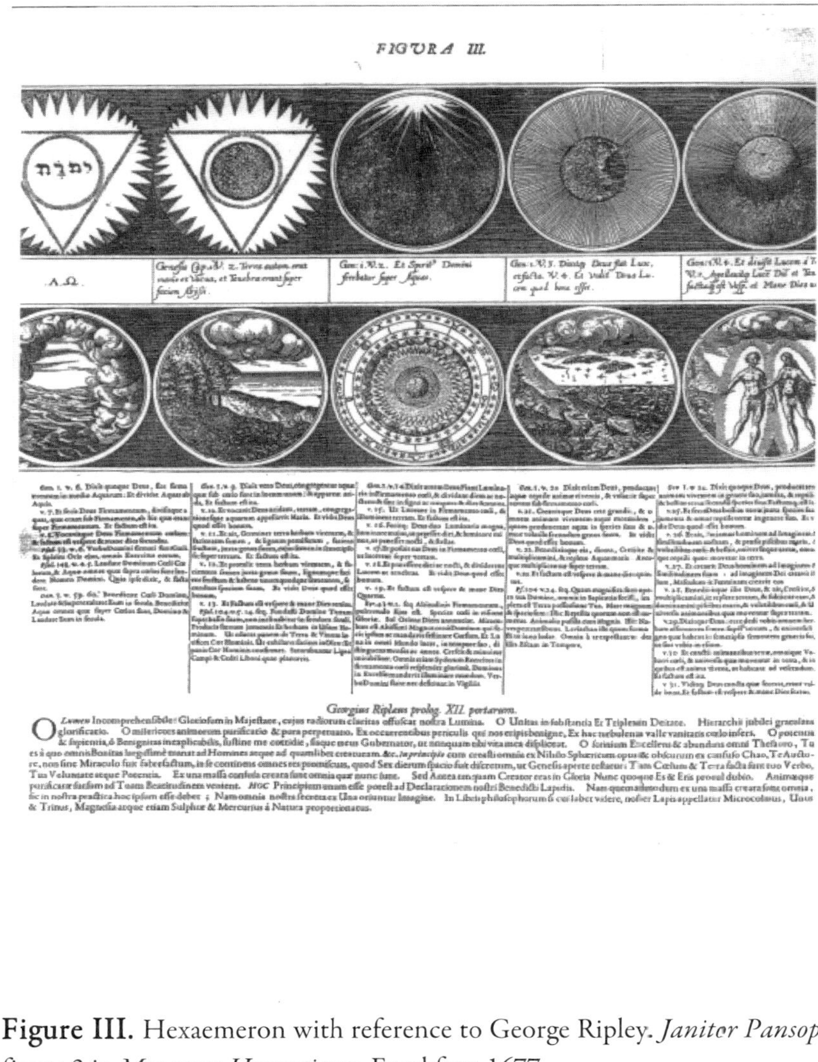

Figure III. Hexaemeron with reference to George Ripley. *Janitor Pansophus*, figure 3 in *Musaeum Hermeticum*. Frankfurt, 1677.

3. The Beginning of the Chemical Reformation: Paracelsus and His Disciples

Paracelsus (Theophrastus Bombastus von Hohenheim) (1493–1541) (*fig. IV*) was "modern" in that he accepted the Hermetic, Platonic, Kabbalistic *prisca theologia* or *doctrina*, which was so much a part of Pico and Ficino's thought.[1] It seems likely, but cannot be definitely shown, that Ficino influenced Paracelsus's work *On Long Life*; in fact, it is very difficult to trace the specific sources of Paracelsus's thought. Some historians connect Paracelsus with medieval authors, such as Arnald de Villanova, Raymond Lull, John of Rupecissa, and John Scotus Erigena, as well as with Renaissance figures, such as Pico, Ficino, Trithemius, and Heinrich Cornelius Agrippa.[2] Yet, Paracelsus's creation of an original "chemical philosophy," through mixing ideas and creating words for his basic theories, truly does not allow a ready demonstration of sources. As Walter Pagel

---

1 Debus in *The Chemical Philosophy* chronicles the history of the chemical philosophy from Paracelsus to the end of the seventeenth century. Debus's work is of such a nature that most studies of chemistry, including this one, expand on it.

2 Walter Pagel, "Paracelsus and the Neoplatonic and Gnostic Tradition," *Ambix* 8 (1960): 125–66.

Figure IV. Paracelsus (Theophrastus Bombastus von Hohenheim).

wrote, "No chemical or alchemical predecessor of the Paracelsian scheme is . . . readily demonstrable."[3]

Paracelsus used his unfamiliar words and concepts with sufficient frequency and clarity that we can identify consistent doctrines. These doctrines appeared sufficiently unequivocal to Paracelsus's spiritual heirs or disciples to enable them to develop a coherent philosophy of nature, related to, but distinguishable from, mere alchemy. Indeed, it seems to have been Paracelsus's goal to refine alchemy, purifying it of its pagan influences, and to perfect it as a revelatory tradition dating from Adam. Rather than considering it an art used to transmute metals, Paracelsus conceived of a refined alchemy as the true Mosaical philosophy, which would be used to heal the sick. He saw himself as a Christian restorer of philosophy. He stated, "I write as a Christian and I am not a heathen, a German not an Italian, an interpreter not a sophist."[4]

I

The Christian alchemical tradition tended to see the alchemical processes, especially distillation, as analogous to the creation processes described in Genesis. Pagan-Gnostic creation-alchemical texts were thought of as expressing truths hidden in Genesis and therefore harmonious with it. Paracelsus espoused a more revolutionary view of the relationship of Genesis to chemistry. In the spirit that moved Luther to reform the church, Paracelsus called for the rejection of the pagan ideas informing alchemy and for the construction of a true Christian chemical art.[5] He did not look to analogies in order to harmonize

---

3 Ibid., 154.

4 "Ich schreib christenlich und bin kein heide, ein Teuscher nicht ein Welscher, ein interpres und nit ein sophist." *Theophrast von Hohenheim genannt Paracelsus: Sämtliche Werke*, ed. Karl Sudhoff, I Abt., 14 vols. (Berlin and Munich: Oldenbourg, 1923–33), 12:10, translation mine.

5 The religious dimension of Paracelsianism is seen in Philipp Homagius's and Georg Zimmermann's Paracelsian antipaganism. See Bruce T. Moran, "Paracelsus, Religion, and Dissent: The Case of Philipp Homagius and Georg Zimmermann," *Ambix* 43 (1996): 65–79. Alchemy also had a significant religious component in the early modern period, as is shown in the work of Stanton J. Linden, "Alchemy and Eschatology in Seventeenth Century Poetry," *Ambix* 31 (1984): 102–24. Religion and knowledge of nature in early modern Germany is outlined in Kathleen Crowther-Heyck, "Wonderful Secrets of

Genesis and chemical processes; rather, he believed that Genesis had to be the source from which the theory of matter and the understanding of chemical processes were derived. Paracelsus saw himself as the chosen one, who freed the true chemistry from the dross that had come to encrust it. His followers saw him in the same way and rejoiced that the chemistry, known to the ancients from Adam to Hermes, was finally available to help mankind. Because he rejected traditional theories, Paracelsus had to create new words to differentiate his Mosaical chemistry from the corrupt semipagan tradition that preceded it.

Paracelsus was regarded by his disciples as a chemical Luther. This is clearly seen in their writings. R. Bostocke, an English Paracelsian of the late sixteenth century, recounted the genealogy of chemistry and the role of Paracelsus in its restoration.[6] Chemistry began with a revelation to Adam, a fact proven by Adam's knowledge of the nature of plants and animals.

> For the Authors, Inventers, originall succession and progression of this auncient Chimycall Phisicke, whose studie and use doeth flowe out of the fountaines of nature, and is collected out of the Mathematicall, naturall, and supernaturall precepts . . . it is to bee understanded, that Adam by divine revelation, or by arte given to him of God. . . . assoone as he did behold any beast [he] did so exactly know all their natures, powers, properties and vertues, that he gave them names, apt, meete and agreeable to their natures and properties . . . he knewe the natures and properties of things better than we, whẽ we have spent all

Nature: Natural Knowledge and Religious Piety in Reformation Germany," *Isis* 94 (2003): 253–73.

6 David Harley argues that Bostocke's first name was Rychard in "Rychard Bostock of Tandridge, Surrey (ca. 1530–1605), M.P., Paracelsian Propagandist and Friend of John Dee," *Ambix* 47 (2000): 29–36. The issue of Bostocke's identity and of whether his name was "Rychard" or "Robert" remains clouded, for William Cooper, the London bookseller, used the name "Robert" in his catalog of chemical books (1673, 1675). Cooper presumably had some knowledge of the first name of the author of *The Difference Betwene the Auncient Phisicke . . . and the Latter Phisicke* (1585). See Stanton J. Linden, ed. *William Cooper's A Catalogue of Chymicall Books, 1673–88* (New York: Garland, 1987), 16.

> our life time in searching out their natures. which was a singuler gift of God & pleased him mightely.[7]

Bostocke cites Josephus for the information that after the Flood worthy men were also educated in the ancient knowledge:

> The sonnes of *Seth* which were his Nephewes, receyving wisdome & knowledge from the hands and delivery of their Auncitors, least that the Mathematicals and that knowledge they had to learned would perish with the fludde, did erect two Pillers, in which they did ingrave their learning, knowledge and invētions, . . . Iosepus writeth in his first booke *Chap. 13*. . . .[8]

Abraham and Moses were keepers of the secrets. Abraham taught them in Egypt and Moses learned them there:

> This *Abraham* having knowledge in the *Mathematicalls*, which in his Countrey were preserved, as is aforesayd by the wonderfull harmony of the worlde did ascende to the knowledge of the one onely God, as some say. But rather, thereby he did see and perceive, the invisible thinges of GOD, that is, his eternall power and Godhead, by the creation of the world. . . .
>
> . . . Sainct *Augustine* sayth in his Booke, *de Civitate dei lib 18. cap. 39.* that the wisedome of the *Aegiptians* wherein Moses was instructed was *Astronomy*, *Geometry*, and *Arithmetike*, and surely it was that which before they had learned of *Abraham*.[9]

The great chain of knowledge, which passed from Abraham to the Egyptians and then back to Moses, reverted again to the Egyptian Toth, also known as Mercurius or Hermes Trismegistus, putative author of the *Aesclapius*. This conflation of wise pagans with God's chosen was a central idea in Renaissance Platonism and Pythagoreanism. This point of view had been embraced by Augustine in the fourth century:

---

7 Allen G. Debus, "An Elizabethan History of Medical Chemistry," *Annals of Science* 18 (1962), reproduces and analyzes part of Bostocke's *The Difference Betwene the Auncient Physicke*. . . . The quotation is found at Debus, 5.

8 Ibid., 5–6.

9 Ibid., 6.

> After these lived *Hermes*, or *Mercurius Trismegestus* in *Aegipt* not long after *Moses*: For alas ye great *Astronomer* lived in *Moses* his time, & was grandfather on the mothers side to *Mercury* the great, whose Nephewes sonne was this *Hermes Trismegestus*, as S. *Augustine* writeth in his Booke, *de Civitate dei lib. 8 Cap. 26.* and *lib. 18 Cap. 39.*... He did write divers workes, as one Dialogue in whiche *Asclapeus* is brought in to thinke that one onely God is the maker of all thinges, and confesseth the errors of them which founde out the superstition of Idols, & doth prophesie in that Booke that they shall perishe.[10]

Despite the efforts of the ancients to hand down and preserve their knowledge of the *prisca doctrina*, the original knowledge became increasingly corrupt. It was necessary for a man dedicated to searching out her secrets to restore God's revelation regarding the workings of nature. This man would not be fooled by pagan error; he would present nature in a manner intelligible to everyone. He would restore the true wisdom, as had John Wycliff, not author a new heresy:

> *Theophrastus Paracelsus*, was not the inventor of this Arte, but the restorer thereof to his puritie.... He was not the author and inventour of this arte as the followers of the Ethnickes [pagans] phisicke doe imagine, as by the former writers may appeare, no more then Wicklife.[11]

Bostocke derived his history from the Paracelsian tradition. Paracelsus recounted a similar history in the *Aurora*.[12] After tracing natural philosophical knowledge from Adam to Abraham, and from the vulcanic Tubal Cain to Moses and the prophets, Paracelsus discussed the differences between the Greeks, the Egyptians, and the Persians. Hermes, an Egyptian, and the Persians more or less adhered to the truth, but the Greeks, as superstitious idolaters, corrupted the truth. Paracelsus himself claimed that he had been called by God to expel Greek-inspired idolatry and superstition from the study of nature. The study of nature did not require pagan sources, for it flourished "with the

---

10 Ibid., 7.

11 Ibid., 25. Bostocke and his English readers probably were more sympathetic to English reformers such as Wycliff than to Germans such as Luther.

12 *The Hermetic and Alchemical Writings of Paracelsus*, trans. Arthur E. Waite, (Berkeley, CA: Shambhala, 1976; originally published London: James Elliot, 1894), 1:48–71.

Germans and other nations, right down to the present day."[13] He, himself, was God's agent, sent to reveal the correct philosophy of Moses:

> The Art having been found out by the superstitious and idolatrous Greeks, some of them who were wiser than the rest betook themselves to the Chaldeans and Egyptians, so that they might draw the same wisdom from their schools. Since, however, the theological study of the law of Moses did not satisfy them, they trusted to their own peculiar genius, and fell away from the right foundation of those natural secrets and arts. This is evident from their fabulous conceptions, and from their errors respecting the doctrine of Moses.[14]

Paracelsus understood that the "doctrine of Moses" contained the correct theory of matter and that from it one could derive the proper use of chemical processes. The Mosaical philosophy led to both a proper study of nature and a correct reading of Genesis. The Christian chemist could legitimately refer to the Gnostic-hermetic corpus because it contained ancient truths descended from Moses and recorded by an Egyptian adept.

II

Paracelsus regarded the creation as a formative movement beginning with God's unity and continuing in the cosmic natural multiplicity.[15] Preexisting with God was a kind of extension or "prime matter." (Paracelsus used different names for this prime matter: *mysterium magnum*, *iliaster*, and *ultima materia*). This mother of all things was prepared by God, but was not created. From it, God separated out, or created, all other things. The act of creation consisted of differentiating individuals or specifics from this divine indefinable, insubstantial material. The first creations were the elements. The

---

13 Ibid., 1:50.

14 Ibid., 1:49–50.

15 Pagel, "Paracelsus and the Neoplatonic and Gnostic Tradition," 144. For the Bible's influence on Paracelsus, see Kurt Goldammer, *Paracelsus in neuen Horizonten: Gesammelte Aufsätze* (Wien: Verband der wissenschaftlichen Gesellschaften Österreichs, Verlag, 1986), and Dane T. Daniel, "Invisible Wombs: Rethinking Paracelsus's Concept of Body and Matter," *Ambix* 53 (2006): 129–42.

means of individuation were the *arcane*, or indwelling forms, which worked on the elements that had been created from prime matter in order to make material things. The separation of objects from the *mysterium magnum* at the birth of the cosmos was "the greatest miracle of philosophy."[16]

Paracelsus believed that his account of God's separation of the multiplicity of the cosmos from preexisting unity was the correct reading of Genesis. In *Das Buch Mineralibus*, he identified ultimate matter (prime matter) with primordial water (not common water), which was with God in the beginning. This water was coterminus with the *mysterium magnum* or *iliaster* from which the elements were created by God's movement across the water in Genesis 1.

III

Paracelsus's chemical interpretation of Genesis became an indispensable aspect of the chemical philosophy. The chemical philosophy was rooted in the belief that it was not a new philosophy, but a return to the original revelation seen in the first book of Moses. The followers of Paracelsus, these new chemists, conveyed their assurance that Paracelsus's reading of Genesis had helped him to purify chemistry from pagan error. The early systematizer of Paracelsus's ideas, Peter Severinus, relied on a chemical interpretation of the creation in his *Idea medicinae philosophicae* (*fig. V*). Severinus believed that Paracelsus had been a genuine Mosaical philosopher and that this fact refuted any anti-Paracelsian charge of heresy. He adopted the Paracelsian reading of Genesis: The Creator had formed all by word and spirit; prime matter was shaped by seeds. "Into these four incorporeal, empty, and vacuous substances, the Creator placed light and the seminal reasons for all things, by an incomprehensible magic, by virtue of that Word and spirit that was moved over the waters, imparting the principles of bodies, in which those things about to come forth onto the world stage were clothed."[17]

---

16 Paracelsus, *Writings*, 2:390–93. See also Pagel, "Paracelsus and the Neoplatonic and Gnostic Tradition," 143, and the chemical dictionary in J[ohn] F[rench], *New Light of Alchemy* (London, 1674), 305–51.

17 Peter Severinus, *Idea medicinae philosophicae* (Basel, 1571), 41–42: "*In his quatuor Naturis incorporeis, inanibus, vacuis, Lucem & seminales rerum omnium Rationes, incomprehensibili*

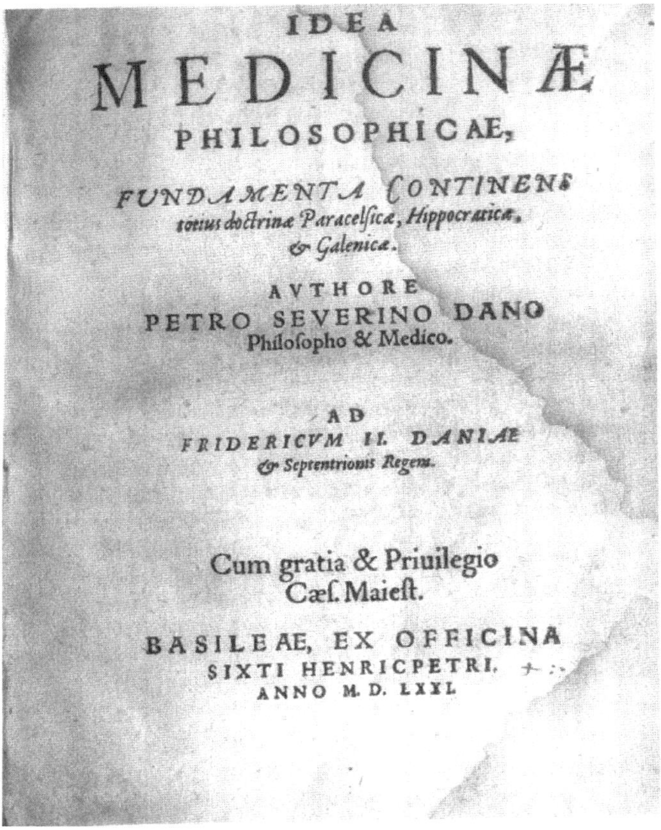

IDEA
MEDICINÆ
PHILOSOPHICAE,
FUNDAMENTA CONTINENS
totius doctrinæ Paracelsicæ, Hippocraticæ,
& Galenicæ.
AVTHORE
PETRO SEVERINO DANO
Philosopho & Medico.
AD
FRIDERICVM II. DANIAE
& Septentrionis Regem.
Cum gratia & Priuilegio
Cæs. Maiest.
BASILEAE, EX OFFICINA
SIXTI HENRICPETRI.
ANNO M. D. LXXI.

Figure V. Title page. Peter Severinus, *Idea medicinae philosophicae*. Basel, 1571.

Magis imposuit Creator, virtute Verbi & Spiritus illius, qui super aquas ferebatur, Principia corporum adiungens, quibus indverentur in mundanam scenam proditura." The translation is from Jole Shackelford, *A Philosophical Path for Paracelsian Medicine* (Copenhagen: Museum Tusculanum Press, 2004), 218.

Thomas Erastus attacked Paracelsianism because of its magical reading of Genesis.[18] Erastus saw both heresy and irrationality in Severinus's belief in a Divine magic that created Paracelsian seeds. He mocked Severinus's and Paracelsus's view of the creation in his *De medicina nova*:

> . . . where they [the followers of Paracelsus] talk nonsense about heaven or fire, air, water, and earth, they write thus: Into these four incorporeal, empty, and vacuous substances, the Creator placed the seminal reasons for all things, by an incomprehensible magic (loving piety more than wisdom, he said, lest he seem to support profane magic, for which Paracelsus, it is agreed, was most eager). . . .[19]

The Paracelsian departure from traditional elemental theory in favor of a magical, chemical understanding of nature and of holy writ made the entire chemical philosophy suspect in Erastus's eyes. He correctly saw that the Paracelsians believed that theirs was *the* Mosaical philosophy. He therefore attacked it as a superstitious perversion of Moses.

The most significant chemist to criticize Paracelsus and his followers was Andreas Libavius. A professor of rhetoric, Libavius was also a serious chemist who wrote many technical treatises. His famous chemical work, *Alchemia*, derides Paracelsians, calling them liars and windbags. His theological criticism of Paracelsianism was remarkably like that of Erastus. He

---

18 Charles D. Gunnoe Jr. has written extensively on Erastus and his criticism of Paracelsianism. Two of his most recent works are "Erastus and Paracelsianism: Theological Motifs in Thomas Erastus' Rejection of Paracelsian Natural Philosophy" in *Reading the Book of Nature*, ed. Allen G. Debus and Michael T. Walton (Kirksville, MO: Sixteenth Century Journal Publishers, 1998), 45–66, and "Paracelsus's Biography Among His Detractors" in *Paracelsian Moments*, ed. Gerhild Scholz Williams and Charles D. Gunnoe Jr. (Kirksville, MO: Truman State University Press, 2002), 3–18. D. P. Walker discusses Erastus's criticism of Paracelsian misreading of Genesis in *Spiritual and Demonic Magic from Ficino to Campanella* (London: University of London, 1958); repr. (Phoenix Mill, Eng.: Sutton, 2000), 156–66.

19 Thomas Erastus, *Disputationum de medicina nova . . . pars altera* (Basel, 1572), 104: "Ubi de coelo vel igne aere, aqua, terra, nagantur, sic scribunt. *In his quatuor naturis incorporeis, inanibus, vacuis, seminales omnium rerum rationes incomprehensibili magis* (amans peitatis dixisset potius, Sapientia; ne sceleratæ videretur favere: cuius Paracelsum studiosissimum fuisse cognostat) *imposiut creator.*" (trans. Shackelford, *A Philosophical Path for Paracelsian Medicine*, 218).

was deeply troubled by Paracelsus's religious heterodoxy and by his reliance on magic and mysticism in his chemical theories. In his *Appendix Necessaria Syntagmatis Arcanorum Chymicorum*, Libavius charged Paracelsians with doing "impure magic" and mixing medicine with theology.[20]

Libavius was interested in both theology and chemistry, but dealt with the subjects separately. When he wrote a history of the "Hexäemeri Mosaici," a theological treatise, he attacked the Paracelsian chemical commentary on Genesis. He took Paracelsus to task for writing in the *Archidoxus* that sulphur, salt, and mercury were created from Hyle, prime matter, by magic. He opposed the notion that when "God said 'fiat,' the word changed into prime matter." It was ridiculous that "fiat" became a sensible thing. Such a notion contradicted the plain language of the Bible, where it is written, "In the beginning, God made. . . ." Libavius believed that Hermes and Paracelsus "produced more a delirium than a sacred physics."[21]

---

20 Andreas Libavius, *Appendix Necessaria Syntagmatis Arconorum Chymicorum* (Frankfurt, 1615), fol. A2: "Sed parcelsici hoc genus nimium dilatant, et *magiae impurae* praetexut faciunt. Ex quinque igitur generibus medicorum *Deale* relagamus ad theologiam."

21 I thank Bruce T. Moran, our contemporary expert on Libavius, for providing me with parts of the text of Libavius's *De Universitate . . . iuxta Historiam Hexäemeri Mosaici in generi Propositam Instituta* (Frankfurt, 1610). The relevent passages follow:

> . . . Paracelsus in Archidoxi magica, non luce solis, sed ignis infernalis digna, scribit Hyliastrum esse primam materiam ex qua sulphur, sal et Mercurium sint creata: ex quo [inte?]lligamus quomdo verbum "fiat" factum sit materiale, tang[i]bile et corpus, in quo omnia sint praedestinata et occultata, quae porro ex Hyliastro per Cogastrum necrolice, necromantice, et necrocomice nascanur, portenta verborum auditis. Sensus videtur esse, cum Deus dixit "fiat", hoc verbum mutatum esse in materiam primam, quam ex [Greek] vocat Hyliastrum, quasi daemon[em?] seu Spiritum illius materiae: (76–77)

and

> Ridiculum vero est verbum "fiat" abiise in corpus illud materiale, sensibile. Cur non legitur statim in principio cum Deus coelum, terram, aerem, aquam, et tenebras facere? . . . Intermiscetur creati`oni lucis paradoxum nouum de seminibus rerum prima die creatis. Hoc unde? Ex Hermeticis et Paracelsicis deliramentis potius quam ex sacra physica. (78, 103).

For Moran's detailed study of Libavius, see his *Andreas Libavius and the Transformation of Alchemy* (Sagamore Beach, MA: Science History Publishers, 2007). Frances A. Yates in *The*

Erastus's and Libavius's attacks on Paracelsus serve to highlight the theological dimension of the chemical philosophy, especially the constant invocation of Genesis by its adherents. Paracelsus bequeathed his followers a disorganized oeuvre, largely in the form of manuscripts strewn throughout the Germanies. With the editing and publishing of the various texts, the disparate concepts were shown to comprise a coherent philosophy. When published, the editors' introductory materials presented the master's systematized Mosaical-Christian chemical philosophy. Among the crucial early Paracelsian editors was Adam von Bodenstein (1528–77), a Swiss physician. His "Epistle Dedicatory" to Paracelsus's *De Vita Longa* is a prime example of how Paracelsian ideas were systematized and placed in a theological context in order to have them accepted as the true Christian science.[22] He begins by stating, "Jehova had shown all creations to be alike in goodness."[23] His identification of Jehovah with the Word in the creation story in the Gospel of John illustrates how Paracelsians read Genesis in light of Christian scripture. (This is also seen in Severinus's discussion quoted above.) If the entire creation is good, then there is a unity of the individual parts of the world with that which is immortal and stable.[24] This unity was illuminated by "doctor Theophrastus Paracelsus."[25] Von Bodenstein argues that Paracelsus's critics did not understand that the doctor had built on an Aristotelian foundation. Moreover his doctrines are clearer than those of Aristotle, for they show what and how God had created.[26] Chemically, Paracelsus demonstrated the existence of the principles of sulphur, oil, and mercury, which are found in all bodies. These principles, according to Von Bodenstein, act in bodies in a geometrical rather than an arithmetical way. This statement of Paracelsian doctrine seems more Pythagorean and Platonic than Aristotelian.[27]

Rosicrucian Enlightenment (1972; London: Routledge Classics, 2002), 69–71, discusses Libavius's objections to the Renaissance mysticism of Paracelsus and his followers.

22 The text is printed in Wilhelm Kühlmann and Joachim Telle, eds., *Der Frühparacelsismus* (Tübingen: Niemeyer, 2001), 211–25. All translations are mine.

23 Ibid., 211, ℓℓ 7–8.

24 Ibid., 212, ℓℓ 21–25.

25 Ibid., ℓ 24.

26 Ibid., 213, ℓℓ 71–98.

27 Ibid., ℓℓ 118–36.

Von Bodenstein cites classical authors to prove that Paracelsus was not revolting against traditional academic knowledge. Paracelsus accepted the Platonic idea that the world was made of prime matter and form.[28] The real issue raised by Paracelsus was how the pagan philosophers could be corrected and their works be made useful in light of the scriptures and the observation of nature. According to Von Bodenstein, Paracelsus had discovered the best way to do this. He had articulated the concepts that everything arises from chemical processes and principles and that God had created these processes and principles from nothing:

> Moreover, God created all according to his divine will through Jesus Christ, the foundation of good, without whom nothing is created. Indeed, the principles were completely created at the beginning of the world . . . so the command of God brought forth the mercurial water in which more salt and sulpher are found than in pure water . . . the principles neither pass away nor are they created but rather remain from the creation of the world.[29]

Von Bodenstein presents Paracelsus as a chemical philosopher-theologian who had provided an explanation of nature that was rooted in a proper understanding of the creation. His doctrines encompass the truths of Pythagoras, Plato, Aristotle, Moses, and St. John. Paracelsian matter theory was the key not only to unlocking the healing powers of nature, but also to understanding the Book of Scripture. This aspect of Paracelsus's thought became an essential element in the texts written by those who studied the treatises of the master, printed and introduced by Von Bodenstein and his brethren of the early Paracelsian fraternity.[30]

IV

Peter Severinus's *Idea medicinae philosophicae* develops an argument similar to Von Bodenstein's argument for the synthesis of Paracelsus with accepted authorities. He "demonstrates" that Paracelsus had restored the care, observa-

---

28 Ibid., ℓ 116.

29 Ibid., 219, ℓℓ 333–46.

30 Kühlmann and Telle discuss the various editors of Paracelsus's texts throughout *Der Frühparacelsismus*.

tion, and commonsense analysis of Hippocrates and Galen to medicine. The new Paracelsian medicine, however, was better suited to treat diseases that had developed since ancient times. Even more important, as noted above, Severinus viewed Paracelsus as a Christian philosopher who had presented ideas derived from holy writ.[31]

Severinus's discussion of theology was brief and superficial compared to the Paracelsian hexaemeron composed by Gerhard Dorn. Dorn, who lived in Frankfurt in the mid-sixteenth century, was an early collector and translator of Paracelsus's works.[32] His *Liber de Naturae luce Physica, ex Genesi desumpta* appeared in 1583. It contains his hexaemeron *Physica Genesis* along with his *Physica Hermetis Trismegisti, Physica Trithemii, Philosophia meditativa*, and *Philosophia chemica*. The full title of the hexaemeron is translated as *The physical light of nature taken from Genesis joined with the wisdom of Theophrastus Paracelsus*. At the outset, Dorn notes that the principle "Arcanum" of all natural philosophy is that God created the world by the operation of light, which "we know today to be Jesus Christ." This fundamental theological insight is connected to the chemical reading of the creation in Paracelsus's *Philosophy to the Athenians* (a work Dorn believed to be authentic; current scholarship questions the attribution to Paracelsus): "The Mysterium magnum brought forth the matrix and the prime matter of all natural creatures . . . all will return to its first essence but not to its prime matter nor to the mysterium magnum."[33]

---

31 See Shackelford's *A Philosophical Path for Paracelsian Medicine* for Severinus's life and works.

32 What little is known about Dorn and his publications is found in J. R. Partington, *A History of Chemistry*, 4 vols. (London: Macmillan, 1961), 2:158–61. Nicholas H. Clulee, "John Dee and the Paracelsians," in *Reading the Book of Nature*, ed. Debus and Walton, discusses Dorn's *Chymisticum Artificium* (1568) at length, and notes that "Gerhard Dorn is less studied and less well known than he ought to be" (119). There are two important studies of Dorn in French: Jean-Francois Marquet, "Philosophie et alchimie chez Gérard Dorn," in *Alchimie et philosophie à la renaissance*, ed. Jean-Claude Margolin and Sylvain Matton (Paris: J. Vrin, 1993), 215–21, and Didier Kahn, "Les débuts de Gérard Dorn d'après le manuscript autographe de sa *Clavis totius Philosophiae Chymisticae* (1565)," in *Analecta Paracelsia*, ed. Joachim Telle (Stuttgart: Franz Steiner Verlag, 1994), 59–126. For an understanding of Dorn and his context, Allen G. Debus's *The Chemical Philosophy* is unexcelled.

33 Argumentum Totius opusculi: Summum atque praecipuum arcanum totius Philosophiae naturalis, in unico immensitatis aternae, entro fundatum, priscos non considerasse, Physicos mirum est. Fortasse quod manifestoribus intehii, ea quae penitus abdita sunt, praterierint, vel potius datum non sit eis,

According to Dorn, Paracelsus taught, as did Moses in Genesis, that water is the fundamental principle of physics. Using Paracelsus's views about prime matter and God's action, Dorn set out to explain the six days of creation.

Dorn believed that the six days were not of the same nature and that the Genesis account dealt with more than one creation. Like Pico, who saw the creation as taking place on three different levels (i.e., the divine realm, the celestial, and the sublunary), Dorn distinguished between two creations, that of the "intelligible world" and that of the "visible world." The intelligible world was created on Day One. From "In the beginning" to the division of light and darkness equaled "one day." The other five days dealt with the visible creation. The creation of the intelligible world comprises the longest part of Dorn's text. There he weaves Genesis with Paracelsus to create a chemical theology of the creation.

Dorn wrote that the creation proceeded from God's utterance and that the idolaters who deny a creator are blasphemers. "In the beginning" indicated to Dorn that the eternal word of God, the Logos of St. John, created the heavens and the earth.[34] Dorn believed that Paracelsus discusses exactly how this had happened in the *Philosophy to the Athenians*. The Word created all things from one matter, the *mysterium magnum*, which is imperceptible and without any essence or form. The *mysterium magnum* is God's greatest work, the substrate of all existence ("ab altissimo opifice Deo"). According to Dorn, Paracelsus deduced his knowledge of the *mysterium magnum* from the "narration of Genesis from God faithfully transmitted by Moses."[35] Indeed,

---

> Deum a suis operibus nosse luclentius. Quod quidem hisce postermis temporibus revelare suis Iesus Christus voluit. De quo etiam primus omnium (quod sciam) Paracelsus libro ad Athenienses, vel ad istorum discipulos scribens, sub involucro docet, mysterium magnum in creatum esse matrem & primam materiam creaturarum omnium naturalium. Et quia dignatum est sese unire quandoque; transiturarum rerum in universum etiam unico centro creato. Cuncta sunt (inquit) in primam suam essentiam redituta, non tamen in suam primum materiam, vel in mysterium magnum. (Gerhard Dorn, *Naturae luce Physica*, in *Theatrum Chemicum*, 6 vols. [Strasbourg: Zetzner, 1659], 1:326)

34 "Nam ve[r]bum Dei . . . Idem sane apparet . . . quod Divus Joan. cap 1. simil sententiam" (ibid., 1:226).

35 "Paracelsus autem ex narratione Geneseos à Deo per Moysen tradita fidelibus . . ." (ibid., 1:338).

from the "uncreated *mysterium magnum*, God created the heavens and the earth";[36] "Moses did not say that the first of creation [*In the beginning*] was the creation, but rather *in it* was the creation of all things."[37]

Moses's words, "God created heaven and earth" do not tell us from what matter they were created. This creation is essentially a unity in which individual things appear and then return to their source. The sentence, "The earth was empty and void," refers to the prime matter from which the elements arose. It also refers to the fluids that are the source of the creatures necessary to human life. Furthermore, the words "darkness on the face of the deep" imply that there is a prime matter of the world. The deep entered into darkness, heaven, and earth, mixing them together and forming one mass.

Moses in his account of the creation indicated that prime matter, heaven, earth, and the complete abyss are water. The spirit of God hovered over this prime matter, the deep and water, which were one, and then divided them. Light pierced the darkness. The emptiness and void were illuminated. First came the heavenly fire, then the elements and the ideal world with its potential. The division of the waters above from those below meant the division of the heaven from the earth. That division is like the division of a woman from a man, when both desire union. This division of powerfully attracted opposites completed the intelligible creation, which was derived from the *mysterium magnum*.

The second day began the visible creation. The creation of the firmament ("fiat firmamentum") divided the heavens from the elements. This was the inception of the visible or physical world. Chemistry is the principle means of understanding the physical world. God, by means of the art of chemistry, divided and delimited the heavens from the elements ("antea diximus coeli divisionem ab elementis arte chemicae dediffinitium"). Paracelsus tried to imitate God's creation by making chemical divisions. Dorn wrote that with his understanding of the creation and the three principles (mercury, sulphur, and salt) Paracelsus could harness natural processes to make beneficial medicines.

Dorn argues that the Genesis account of the other days of creation also agrees with the teachings of Paracelsus. Dorn's detailed harmonization of

---

36 "In mysterio magno increato creavit Deus coelum et terram" (ibid.).

37 "Non dicit etiam Moyses creationis principium fuisse creatum sed in eo creationem omnem factam esse asserit." Ibid.

Paracelsus's doctrines with scripture not only places Paracelsianism within the hoary hexaemeric tradition but also casts it into a theological form familiar to sixteenth-century religious orthodoxy.

In the decade after the publication of Dorn's hexaemeron, Oswald Croll (1560–1609) produced his *Basilica Chymica*.[38] It served to further advertise and establish the chemical philosophy as an important approach to the Christian study of nature. The work, published in Frankfurt in 1609, consists of two parts, the *Admonitory Preface*, a long introduction to Paracelsian thought, and a pharmacopoeia of chemical remedies. A separate treatise on the *Internal Signatures of Things* was printed with the *Basilica*. Croll was an academically trained German physician who continued in Severinus's footsteps as a harmonizer of Paracelsus with traditional medicine. He relied both explicitly and implicitly on Severinus. For example, he begins the pharmacopoeia by paraphrasing Severinus's statement regarding the two types of healing.[39] More to the point of this discussion of Genesis and chemistry is Croll's situating of chemistry in the context of the creation.

Croll presents the chemical philosophy as a tradition that began with the revelation of the creation and continued through the prophets, Hermes, Philo, and St. Paul. He declares that Paracelsus had restored to chemistry a correct understanding of the elements and chemical processes. The chemical reading of Genesis, a reading that reveals nature and the Creator, proved the faithfulness of Paracelsus's restoration. He described how the chemical philosophy agrees with Philo's view of everything flowing from divine wisdom. He further affirms the chemical-philosophical orthodoxy of Severinus's discussion of the earth and the heavens.[40] Chemistry proved that God is the foundation of nature and, hence, of medicine.

For Croll, the most fundamental proof of the correctness of the chemical philosophy is its congruence with Genesis. God, by means of the word (*fiat*) created prime matter. Prime matter contained the three Paracelsian principles or elements: mercury, sulphur, and salt. From these three, all bodies were

---

38 The Latin edition used in this discussion is Oswald Croll, *Basilica Chymica* . . . [published with] *Tractatus novus de Signaturis rerum internis* (Frankfurt, 1611; repr. Hildesheim: George Olms, 1996). Translations mine.

39 Ibid., 11.

40 Ibid., 16.

formed.[41] In the same mystical language of Paracelsus that so upset Erastus and Libavius, Croll wrote that the Creator brought forth multiplicity from unity by means of the "incomprehensible magic" of the "eternal treasury of divine wisdom" acting on prime matter. This was done by the *Word* and the *Spirit* moving across the face of the water. The seeds gathered together at this, the beginning, are still the active principles in the world.[42]

It was from Genesis that the chemical philosophy derived its theory of matter and its methodology for using the creative powers still resident in nature to create healing medicines. The chemist emulated the creation and used the powers of the seeds created by God at the beginning. Just as chemistry learned from Genesis, so Genesis proved chemistry to be the true revealer of nature.

In Croll's view, the persistence of the effects of the creation in the world is seen in the internal signature of things. The signature is the image of God impressed in the individualities that arose from the one. The signatures of plants, animals, and minerals, if properly read, reveal their healing virtues.[43] God in the truest sense fills nature.[44] The world is God evolving.[45] Signatures reveal his presence. This Kabbalah of received knowledge of signatures was revealed first to Adam. Adam saw the true essence of each thing and named it. The Adamic name was actually the thing itself.[46] Humans can know the signatures in nature because the light of God is in them. God's image is in things, and the human mind or intelligence is in God.[47] Our *mens* connects us to God and hence to the inner meaning of things. This is the source of medical knowledge. Such knowledge is possible only when the creation is correctly, that is chemically, understood.

Croll's detailed presentation of Paracelsianism became a major source for aspiring chemical philosophers. He discusses the chemical philosophy as part of the academic tradition, as a hermetic science, and as a Kabbalah that explains Genesis. The *Basilica Chymica* was translated into French, German,

---

41 Ibid., 18.

42 Ibid., 19.

43 Croll, *De Signaturis rerum internis.*

44 Ibid., 3.

45 Ibid., 14.

46 Ibid., 15.

47 Ibid., 14.

and English.[48] It was, perhaps, the most successful Paracelsian primer because it combines theory with a useful pharmacopoeia.

As an essential aspect of Paracelsianism, the reliance on Genesis is also central to Croll's work. It is this religious element that made the chemical philosophy attractive to so many intellectuals in the religiously charged environment of the sixteenth and seventeenth centuries. This is well illustrated in the reception of the chemical philosophy in England.

V

In 1585 R. Bostocke had presented Paracelsus to the English as a medical Wyclif. Shortly thereafter, a clergyman, Thomas Tymme, influenced by its theological content, took up the promulgation of the chemical philosophy. Tymme apparently had Calvinist leanings, at least around 1573 when his translation of Calvin's commentary on the Epistles to the Corinthians, with a dedication to Edmund Grindal, Archbishop of Canterbury, was published.[49] What exactly occasioned his interest in the chemical philosophy is unknown, but he certainly desired to understand the book of scripture through a study of the book of nature. This was, of course, the primary claim that Paracelsus and his followers had made.

Tymme turned from translating Calvin to translating materials from the French Paracelsian Joseph Duchesne, also called Quercetanus. In 1605, Tymme produced *The Practice of Chemicall, and Hermeticall Physicke*.[50] Although the Library of Congress suggests that *The Practice* is a translation of *Ad veritatem hermeticae medicinae ex Hippocratis . . .* , this is not exactly the case. Tymme translated selected parts of *Ad veritatem*, but not all of it. In addition, he "shewed thee such secrets in this Art, as neither Quercetanus, . . . nor any other

---

48 The French translation of *Basilica Chymica* appeared in 1624 and the German in 1629. *De Signaturis* was translated into English in 1669, the "Admonitory Preface" in 1657, and the entire work in 1670. There were as many as eighteen editions from 1609 to 1658.

49 John Calvin, *A Commentary upon Saint Paul's Epistles to the Corinthians*, trans. Thomas Timme (London, 1573).

50 Josephus Quercetanus, *The Practice of Chymicall and Hermeticall Physicke*, trans. Thomas Timme (London, 1605). See Bruce Janacek, "Thomas Tymme and Natural Philosophy: Prophecy, Alchemical Theology and the Book of Nature," *Sixteenth Century Journal* 30 (1999): 987–1007.

Phylosopher, have before published in print to my knowledge ...".[51] This unpublished material was from manuscripts in his possession. The first chapter, which he takes from page 4 and following of *Ad veritatem*, gives a genealogy of chemistry from the Hebrews, through the Chaldeans and Egyptians, to the Arabs. In chapter 2, he relates how the world is said to have been created by God through the word. This truth contradicts Aristotle and other "ethnicks." "Elohyim, or great God," which moved on the waters, is the same as the "World Soul of the Platonist," which enlivens the world. The waters, therefore, are the source of chemical action by means of the divine creative principle. The quickening spirit of "that great Trismegistus" is better described "in Moses writings, then all other Philosophers." The creation, the power of the divine spirit, and the building of a Christian rather than an "ethnick" understanding of nature are central to the theoretical framework of *The Practice.* Understanding this framework is the foundation of all knowledge.

The "Epistle Dedicatory" to *The Practice* is more important for understanding why Tymme became a chemical theologian than the "translated" text assembled by him. Indeed, such introductory materials are the best source for comprehending why English intellectuals chose to study the chemical philosophy. Tymme's Epistle clearly presents "halchymicall" knowledge as an essential adjunct to theology:

> I[t] may seeme (Right Honorable) an admirable and new *Paradox*, that *Halchymie* should have concurrence and antiquitie with *Theologie*, the one seeming meere *Humane*, and the other *Divine*. And yet *Moses*, that auncient Theologue, describing & expressing the most wonderfull Architecture of this great world, tels us that the *Spirit of God moved upon the water*: which was an indigested Chaos or masse created before by God, with confused Earth in mixture yet, by his Halchymicall Extraction, Seperation, Sublimation, and Coniunction, so ordered and conioyned againe, as they are manifestly seene a part and sundered: in Earth, Fyer included, (which is a third Element) and Ayre, (a fourth) in Water, howbeit invisibly. Of which foure Elements, two are fixed, as earth and fire: and two volatil, as water & ayre.[52]

---

51 Quercetanus, *Practice*, final page (text is unpaginated).

52 Ibid., A3.

The action of the Spirit of God brought forth from chaos the Paracelsian *tria prima* (mercury, sulphur, and salt), which, in turn, gave rise to all minerals, vegetables, and animals:

> That spiritual Motion of the first mover, God, hath inspired al the creatures of this universal world, with that spirit of Life (which may truely be called the spirit of the world) which naturally moveth, and secretly acteth in all creatures, giving them existence in three, to wit, salt, sulphure, and Mercury, in one *Hupostasis*. Mercurie congealing Sulphur, & sulphur Mercurie, neither of them being without their Salt, the chiefest meane by whose helpe Nature bringeth forth al vegetals, Minerals, & Animals.[53]

This divine alchemy was the beginning of time and terrestrial existence. "In the end of days," this alchemy will change all and make "New Heavens and a New Earth, and bring all things to a crystalline cleerness," as is described in the Apocalypse of John 21:5. By alchemy, Tymme means Paracelsian chemistry. The chemical philosophy thus became a key not only to understanding nature specifically but also to envisioning the latter days and interpreting the scriptures in general. Tymme wrote,

> This Philosophy natural, both speculative & active, is not only to be found in the volume of nature, but also in the sacred Scripture: as in *Genesis*, in the booke of *Job*, in the *Psalmes*, in *Syrach*, and in other places.
>
> In the knowledge of this Philosophy, God made *Salomon* to excel all the kings & Phylosophers.[54]

Like Bostocke, Tymme believed that the chemical philosophy was a restoration of ancient Hebraic wisdom. Paracelsus's work was not original; nonetheless, he was a tool made by God for the purpose of bringing to mankind illumination with God's original light. This would lead to the discovery of effective medicines but even more importantly to insights into the meaning of scripture. God could be comprehended only through study of both the book of scripture and the book of nature. For Tymme, translating Paracelsian texts was not a rejection of his previous commitment to Calvin and theology,

---

53 Ibid.

54 Ibid., A[5].

but an extension of it. Tymme believed that chemical philosophy was theology. This "fact" was the theme of Tymme's own composition, *A Dialogue Philosophicall*, published in 1612.[55]

Tymme's *Dialogue* is a discussion of "naturall Philosophie" between Theophrast and Philadelph. Theophrast presents the Paracelsian view of matter and natural processes: All is composed of matter and form and exists in the number three. Again, Tymme's "To the Reader" is the focused and succinct statement of why he believed in Paracelsianism. It all came down to the Paracelsian reading of the book of nature, which was a light to understanding God. Paracelsian chemistry taught a thoroughly Christian view of nature, untainted by pagan error:

> The Almighty Creatour of the Heavens and the Earth, (Christian Reader), hath set before our eyes two most principall Bookes: the one of Nature, the other of his written Word. By these we know that God was before all times, infinitely one, comprehending all things that now be, in himselfe: and being the beginning of Time, created those things in number, measure, and waight, adorning his worke most wonderfully with these three Instruments, as with the first Formes.[56]

The study of nature is the contemplation of God. Celestial and earthly motions bespeak his glory:

> The wisedome of Natures booke, men commonly call *Naturall Philosophie*, which serveth to allure to the contemplation of that great and incomprehensible God, that wee might glorifie him in the greatnesse of his worke. For the ruled Motions of the Orbes: the wonderfull workmanship of so many starrie Tents ... he is manifested in them, and by them, as their finall cause to whom also they tend.[57]

The book of nature, properly read, awakens the soul. The Fall of Adam brought corruption to mankind. That corruption keeps the soul from perceiving God in nature. The Mosaical philosophy could remove corruption

---

55 Thomas Tymme, *A Dialogue Philosophicall wherein Natures Secret Closet is Opened ...* (London, 1612).

56 Ibid., A3.

57 Ibid.

and teach the pure soul both natural philosophy and morality. Tymme makes the extraordinary statement that, without the Fall of Man, the holy scriptures would not be necessary, for a soul without original sin would see itself—the image of God—in nature:

> If man had not sinned, the booke of *Nature* would have sufficed to have kept him alwaies in the knowledge & obedience of God his Creator. For then he should himselfe have carried that Booke whole & perfect, imprinted in his heart and minde, neither should his Soule have needed any other Teacher to know it selfe; but it selfe should have clearely beheld and contemplated it selfe so long as she preferred and preserved her first light, wherein God had created her.[58]

Tymme's understanding of Christianity led him to embrace the chemical philosophy and its religious reading of the book of nature. For this same reason, another Englishman, H. Pinnell, translated Croll's *Admonitory Preface* and Paracelsus's *Philosophy to the Athenians* into English in 1657.[59] Pinnell's "Translator's Apology" notes that God is seen in all things and that this is what Croll and Paracelsus teach us. True chemistry, like true religion, finds light in the darkness. While Pinnell does not assert that the book of nature is superior to scripture, he states in his "Postscript" that they establish one another:

> Whereas I have proposed and pressed the Creation of God as an object of thy admiration and motive to his adoration, it is not, in the least, intended any way to diminish the due Reverence rightfully belonging to that more rich mercy of the Lord added to man in giving and preserving those sacred Records of heavenly mysteries contained in the most holy Scriptures; my desire rather is that both those Books of God (Nature and Scripture) might be better studyed and more observed. Doe I seek to make

58 Ibid., [A2].

59 H. Pinnell, ed. and trans., *Philosophy Reformed and Improved, in Four Profound Tractates* (London, 1657).

> voyd the Word of the Lord by his Works? God forbid; nay I establish the one by the other.[60]

The theological basis and hence attraction of Paracelsus and his followers for many intellectuals is beautifully stated by John French, MD, in his English translations of Sandivogius's *New Light of Alchymy* and Paracelsus's *Nine Books of the Nature of Things* in 1674 (*fig. VI*).[61] French believed that the first two chapters of Genesis were the essence of religion and philosophy. It was the chemical philosophy that opened them to seekers in truth and clarity:

> Do not all grant that the two first Chapters of *Gen*. are true Divinity? I dare also affirm that they are the most deep and truest Philosophy. Yea, they are the ground and sum of all Divinity and Philosophy: and if rightly understood, will teach thee more knowedg of God, and thy self, than all the Books in the world besides. Now for the better understanding of them, make use of most profound *Sandivogius*, the Author of the first of the ensuing Treatises, as the best Expositor of them: in that Treatise of his thou shalt see the mystery of the Deity and Nature unfolded, even to admiration: as to see what that Light and Fire is which is the Throne of Gods Majesty. How he is in the Heaven most gloriously, and in the Creatures providentially. How he is in the Life of that universal Spirit which is diffused thorough the whole world. What that Spirit of his is that moved upon the Waters.[62]

For Paracelsians, the chemical study of nature was an act of faith undertaken in the service of God. The chemical philosophy did not remain purely Paracelsian but grew and developed. Its practitioners were often critical of its founder just as later theologians were of Wyclif, Luther, and other reformers. Moreover, chemical philosophers were extraordinarily idiosyncratic and produced very different and nuanced readings of the book of nature, as we shall see in the next chapter. Yet they tended to share a sense of religious

---

60 Ibid., "Postscript" unpaginated.

61 Sandivogius, *New Light of Alchymy* also Paracelsus, *Nine Books of the Nature of Things*, trans. J[ohn] F[rench], M.D. (London, 1674).

62 French, "To the Reader," in *New Light of Alchymy*, by Sandivogius, unpaginated.

calling and a belief that the book of scripture could be understood only if read by the light of the book of nature. As self-aware spiritual heirs of Moses, they took the first chapter of Genesis as the foundation of their chemical study of nature.

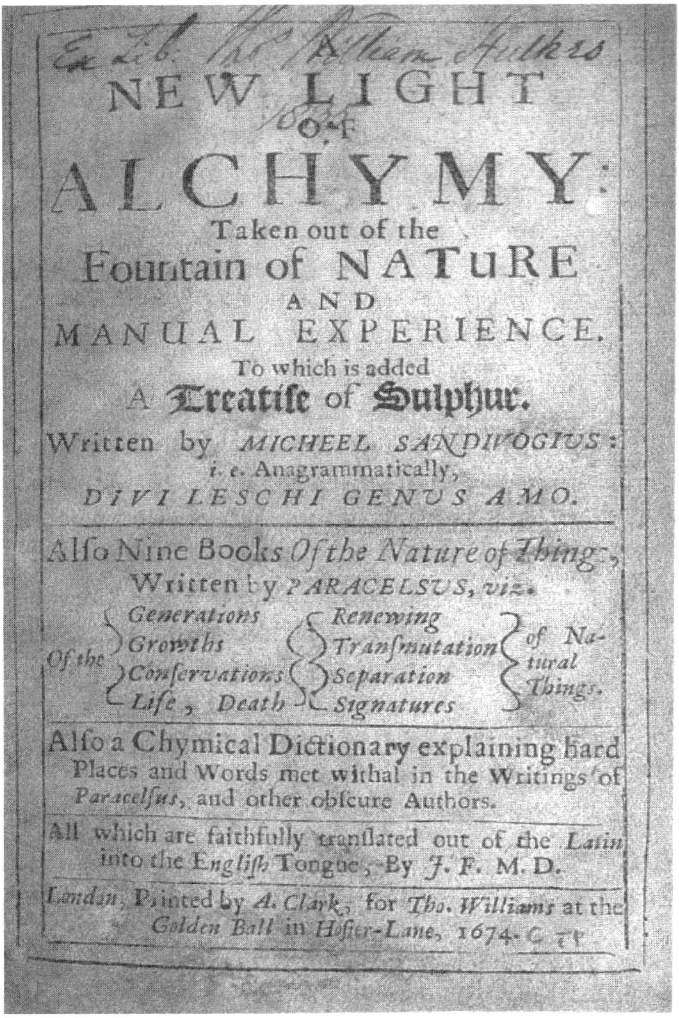

A
NEW LIGHT
OF
ALCHYMY:
Taken out of the
Fountain of NATURE
AND
MANUAL EXPERIENCE.
To which is added
A Treatise of Sulphur.
Written by *MICHEEL SANDIVOGIUS*:
i. e. Anagrammatically,
DIVI LESCHI GENUS AMO.

Also Nine Books *Of the Nature of Things*,
Written by *PARACELSUS*, *viz.*
Of the *Generations*, *Growths*, *Conservations*, *Life*, *Death*, *Renewing*, *Transmutation*, *Separation*, *Signatures* *of Natural Things.*

Also a Chymical Dictionary explaining hard Places and Words met withal in the Writings of *Paracelsus*, and other obscure Authors.

All which are faithfully translated out of the *Latin* into the *English* Tongue, By *J. F.* M. D.

London, Printed by *A. Clark*, for *Tho. Williams* at the *Golden Ball* in *Hosier-Lane*, 1674.

Figure VI. Title page. Micheel Sandivogius [Michael Sendivogius], *New Light of Alchymy*, trans. J[ohn] F[rench], M.D. London, 1674.

4. The Chemical Philosophy Matures: Heinrich Khunrath, Robert Fludd, Joan Baptista Van Helmont, and Johann Rudolf Glauber

The chemical philosophy became more than Paracelsianism. Polymaths such as Heinrich Khunrath and Robert Fludd elaborated chemistry in a complex of Renaissance Neo-Pythagoreanism, Neoplatonism, and Christian Kabbalah. They demonstrated a thematic intellectual unity of the major streams of natural and theological knowledge. This seemed to confirm that all knowledge descended from a *prisca doctrina*, originating in God's revelation to the Hebrews. Their learning was profound and their works reveal a richness and depth, for they are filled with citations to specific ancient sources. Paracelsus did not, and perhaps could not, identify the sources of his ideas, which are often seen only as intimations in his work.

In contrast to Khunrath and Fludd stood Joan Baptista Van Helmont. Van Helmont, too, was profoundly learned and knowledgeable about Renaissance philosophy and Christian Hebraism. He, however, more sharply focused his thought on chemistry and medicine. He lamented the amorphousness and lack of rigor in Paracelsus and many of his followers. Van Helmont acknowledged Paracelsus's insights, but he sought to refine and improve upon them. In the work of Van Helmont, one can see an important development of the Christian chemical philosophy. Van Helmont left behind the bombast and idiosyncrasies of the "master" and focused on relating precise laboratory work to theory. Van Helmont restated the chemical philosophy and differentiated it from Paracelsianism.[1]

Van Helmont reported and interpreted many reproducible experiments that are extremely practical in nature. In this, he can be seen as part of the pragmatic chemical tradition represented by Johann Rudolph Glauber. Glauber presented useful methods for producing industrial chemicals and medicines. He hoped that chemistry would benefit humanity in health, nutrition, and material goods. Glauber, like Van Helmont, Khunrath, and Fludd, however, saw chemistry in a theological light. Genesis was for him, as it was for the others, a text intimately associated with chemistry. Genesis, properly understood, justified the study and use of chemistry. It also provided information that was of great use to chemical theory and practice.

I

Heinrich Khunrath (1560–1605) was a practicing physician who was very well read in classical texts, Renaissance Neoplatonism, chemical texts, and Christian Hebraism. As a devout, mystical Christian, he sought to demonstrate that these traditions, taken together, constituted a coherent worldview. The various strands of knowledge were part of the *prisca doctrina* revealed by God to his chosen several times throughout history. Khunrath believed that he belonged among those inspired by the Holy Spirit, chosen to demonstrate the unity of true wisdom and received traditions, no matter how tarnished those traditions had become in transmission.

Khunrath's most famous work was the *Amphitheatrum sapientiae aeternae,* which brought theology (the book of scripture) together with natural philosophy (the book of nature) in a grand scheme, a true and complete Christian philosophy. He wrote other texts that dealt with aspects of his "catholic" worldview, such as the *Confessio de chao physico-chemicorum catholico*; *Von hylealischen das ist, pri-materialischen catholische, n oder algemeinem natürlichen ... chaos*; *Symbolum physico-chymicum*; and *De igne magorum Philosophorum* whichmentions and summarizes his previous texts,

---

1 Van Helmont criticized Paracelsus from within the chemical philosophical tradition. He, like Paracelsus, wanted a true, Christian science. He is thus distinguishable from Libavius, who wanted a chemistry untainted by heterodox, mystical theology.

explaining the nature of the "great fire" in its external and internal (spiritual) dimensions. Khunrath built his philosophy on his reading of scripture, especially Genesis. I shall trace his thought on the Creation in the *Amphitheatrum* and its distilled essence in *De igne magorum philosophorum*.[2]

Amphitheatrum opens with a schema (painted in words, not pictures) of its organization and contents. On the left side of the page are topics contained within theology, headed by the Unknown God and Jesus Christ, whom He sent. On the right side of the page are topics within the Christian Kabbalah, divine magic, chemistry, and the three all-encompassing things, namely, the microcosm, the macrocosm, and all the good things in this world. The focus of *Amphitheatrum* is the mystery of the Trinity. Khunrath discusses the Trinity in terms of theology, Kabbalah, and chemistry, which are the sum of philosophy in its spiritual and physical dimensions.

Khunrath's philosophy, as a commentary on Genesis, is set out pictorially in plate 2, the hermaphrodite. In the text surrounding the illustration, Khunrath writes that nature is known generally and particularly from the Holy Scriptures.[3] Nature itself—that is, the great world and the lesser world (Man)—proceeds from God, "theosophically, physically, physico-medically, physico-chemically, physico-magically, hyperphysico-magically, [and] Kabbalistically."[4] The philosophers' stone is the essence of the creation because it contains within itself all of the processes found spiritually and physically in nature [5]

---

2 Khunrath's bibliography, listed in the text above, demonstrates an interest in chemistry and medicine very similar to that of Paracelsus. Urszula Szulakowska in "John Dee and European Alchemy," The Durham Thomas Harriot Seminar Occasional Paper, no. 21 (Durham, Eng.: University of Durham, 1996), discusses Khunrath with respect to Paracelsus and Dee. She also notes the difficulty in dating editions of Khunrath's *Amphitheatrum*. Peter J. Forshaw has also written on Dee and Khunrath and is preparing a biography of Khunrath: Forshaw, "The Early Alchemical Reception of John Dee's *Monas Hieroglyphica*," *Ambix* 52 (2005): 247–69, and "Subliming Spirits: Physical-Chemistry and Theo-Alchemy in the Works of Heinrich Khunrath," in *Mystical Metal of Gold*, ed. Stanton J. Linden (New York: AMS Press, 2007).

3 Heinrich Khunrath, *Amphitheatrum sapientiae aeternae* (Hamburg, 1595), plate 2, "Naturam nosce, universaliter, et particulariter, ex libro sanctisimae Scripturae. . . ." Translations are mine.

4 Ibid., "ex Deo Ipso: theosophice; physice; physicomedice; physicochemice; physicomagice; hyperphysicomagice; cabalice."

5 Ibid., ℓ 2.

The philosophers' stone was the means by which the Divine Trinity created the entire world.[6] The stone is nothing other than "the Spirit of God, the *ruach elohim*, which incubated the waters in Genesis I." In the midst of the heavens, the Tetragrammaton flew and made the great world from the chaos. The world was born in the thought of the Son. The hermaphrodite represented the various opposing forces at work at the Creation.[7] These forces turned out to be chemical in nature. Khunrath's understanding of Genesis 1 is consistent throughout his works. The contemporary English translation of his *Confessio de chao* elucidated, more briefly than the *Amphitheatrum*, the divine chemistry of the *ruach elohim:*

> But it is further necessary in the Catholike Chymicall Arte, yt we truly believe (ye know directly) & stedfastly hold that Ruach Elohim the Spirit of ye Lord wch in Gen:1.2. did move upon the Water, is really become a body.
>
> Now therefore this godly Wisdome is true Philoosophy, that we acknowledge & confesse, yt Magnesia of the the greate world, (the greate worlde Sustayner & cherisher) S'aim Ruach Elohim, yt is the Spirit of the Lord or the expression of the Trinity of the God head, a Father, Son & holy ghost, a godly power or emanacion of the godly power or a greate litle world.
>
> That Spirit of the Lord was OR[8] preceeding of the Essence of the Spirit of the Lord (wch in ye beginning of the world was wth the Lord) wch Gen: 1.2 moved upon the waters, wch Wisd: 1,7 filled the whole world, and wch Wisd: 11,26 is in all even in ye most inward & sacredest Virgin wombe and center of the Earth, the most misticall broodie Mother of the greate world. world, bodily it is become a Corporall Salt of wisdome, tho litle[,] greate world OR bred in the world, of the bodily essence & first matter yt is earth & water, the Parents, to wittt of ye great World.[9]

---

6 Ibid., ℓ 6.

7 Ibid., "Lapis Philosphrum est *ruach elohim* Ruach Eliohim, (qui incubabat aquis, Gen. 1.)."

8 The word for "light" in Hebrew. Khunrath's idea of light is explored in Urszula Szulakowska, *The Alchemy of Light* (Leiden: Brill, 2000), 79–137.

9 English translation of Khunrath's *Confessio de chao*, British Librabry, Ashmole MS 1459, 99–106; http://www.levity.com/alchemy/khunconf.html.

Khunrath saw the Creation as a theological-chemical moment that is profoundly connected to the proper use of nature to heal the sick and to bring eternal salvation. The Son is the light of creation, the philosophers' stone, and the source of the *ruach elohim*. The Creation is at one level a parable of Jesus's atonement. Khunrath read Genesis 1 in light of the Hebrew scriptures' identification of wisdom as God's creative tool, the Christian Gospel of John's creation by the Word, and the triumphant appearance of Jesus in the Apocalypse of John. Khunrath presented these texts as pure theological-chemistry:

> Which when OR [Jesus] hath suffered Chymically and naturally as aforesaid [yea about ye universall chimicall godly magick & Christian cabalisticall salvacion of man as I have tought in ye 4 figures of my Amphitheatre of Eternal Wisdome], for yt safety of ye greate world & is by natural opperacion buried & descended into Hell, in the third of his oft mencioned labours, or worke, he shall rise from death without impuritie or corruptableness, Ascende into the heaven of his more than full perfection, & in full brightness sitteth at the right power and strength of the great world, his potent father in a Christaline transparentness & Carbuncle like redness, wth out corruptableness compleat in Body, Soule and Spirit, like a perfect Quintessence.[10]

Jesus as Light, the creative spirit, and the essence of the chemical processes is the basis of Khunrath's *De igne magorum philosophorum*. The bulk of the work is written in German, but "To the Reader" is in Latin. In it, Khunrath writes that "light has many forms and its 'Imago' [image-essence] is diffused in all things from the divine mind."[11] To paraphrase Khunrath, God [the Father] is the first—the Father of light. Second, the Son is his [the Father's] splendor and exuberance. Third, the Holy Spirit is effulgence [spiritual insight] above all intelligence. The Trinity, in the form of light, is glimpsed in nature as the "natural fire" situated in the heavens. The German text begins by stating that the philosophers' stone is not prepared with the heat of external or visible fire; rather, it is prepared with the inner nature of

---

10 Ibid. Brackets in original except for [Jesus].

11 Heinrich Khunrath, *De igne magorum philosophorum* (Strassburg, 1608), ad lectorem, 3. Translations are mine.

fire. Fire in the chemical sense is "an artificial God of chemistry."[12] The adept can use this fire correctly *only* with the help of the *ruach elohim.*[13]

Khunrath saw philosophical fire, the light of creation, as an essential aspect of the *prisca* tradition. It was symbolically represented on the altars of temples, including those at Delphi; at the shrine of goddess Vesta in Rome; and, most perfectly, at the temple in Jerusalem. The power of the godly fire possessed by the Jews was demonstrated by the deaths of Nadab and Abihu, the sons of Aaron who brought fire into the tabernacle on their own initiative, rather than by God's command. The fire on the altar in Jerusalem was like the heavenly fire.[14] That fire or light was the soul of the world, or *naturas naturae.* It was truly the form, or soul, that God placed into the "globe of the world."[15]

The goal of philosophy was to understand the creation theologically and physically. In this belief, Khunrath is like the Paracelsians; yet his greater reliance on specific Christian Kabbalistic doctrines such as the *ruach elohim,* Neoplatonic-Kabbalistic ideas of light, and chemistry distinguish his thought as more complex and theosophical than that of Paracelsus's disciples. Khunrath's "salvation" chemistry goes beyond Paracelsianism. With Khunrath, we have Jesus as Light—light as the instrument of creation, and light or fire as the chemical manifestation of God, with the power to make medical remedies. His salvation chemistry characterizes Khunrath as a theosophical chemist, a priest in the tradition of Aaron. To be a chemist, Khunrath confessed, one must know the Or (light, Jesus), who will save and judge the world. Only divine illumination could make the chemist a true Christian philosopher, one who could make the philosophers' stone and heal the sick, for is not Jesus, the healer, the philosophers' stone.[16]

Genesis was seen as the key to theology, to chemistry, and ultimately to salvation:

> From hence cometh OR Compleate by naturall opperacion
> to Judge both the litle and the great world, sound & sick.

---

12 Ibid., 3.

13 Ibid., 4.

14 Ibid., 8.

15 Ibid., 21.

16 English translation of Khunrath's *Confessio de chao*, Ashmole MS 1459, 106.

> And at his coming over ye Elementall offspring shalbe sifted and Judged (as naturally vegetables, animalls, mineralls & mettalls) and purified in their bodyes Spirit & Soules, & must each in their kinde give an account of their good or evill.
>
> But if there be good found in them it goeth into glory, but if Evill into to fire of destrucion.

The secret of the philosophers' stone lay in the proper use of fire on the universal mercury. Only a chemist obedient to God could make the philosophers' stone:

> This is the assured true doctrine of the Wise Concerning the Catholike universall Mercury of the wise Philosophers.
>
> That Chymist wch doth not truly know, steadfastly believe, & firmly hold the same; he cannot be a true naturale Chymicall Philosopher, neither can he ever obteyne yt greate & universall Stone of the philosophers. [Not through wisdome or Sophistry but according to ye wisdomme yt cometh from God, of wch you may see more in ye 2: chap of my Narurall Chymicall Confession in Dutch].[17]

Khunrath did not try to write obscurely; rather, he tried to present spiritual-natural truths clearly in the argot of the art. He believed that the Holy Spirit, the *ruach elohim*, directed his writing. He wrote for the benefit of pious seekers of truth—for their spiritual welfare; for their intellectual growth; and, above all, for their salvation. As he professes,

> I Doctor Kunwrath through Divine wisdomme have written thys the 12 of December about the houre of 8 in the Evening Anno Dm 1597.
>
> Instigated
By the allpowerfull spirit
Praise you the Lord, Praise you the Lord.
Praise you they Lord.[18]

17 Ibid. Brackets in original.

18 Ibid.

II

Robert Fludd (1574–1637) was kindred in spirit to Heinrich Khunrath. His version of the chemical philosophy was likewise a mixture of classical erudition, Christian Hebraism, and the medical-chemical tradition. His many volumes are prized for their lavish illustrations and all-embracing learning that was becoming somewhat old-fashioned by the first quarter of the seventeenth century.[19] The work that most clearly expresses his philosophy appeared posthumously in Latin in 1638 and in English (probably from Fludd's own manuscript) in 1650. The title *Mosaicall Philosophy* clearly indicates that Fludd felt his understanding of nature not only was derived from Moses, but was essentially Moses's philosophy. He believed himself to be in the long tradition of "Mosaicall" philosophers descending from Sinai. In this, he distinguished himself from the followers of the "ethnicks" or pagan philosophers, whom he believed had corrupted Christian thought.

Robert Fludd saw himself as the explicator of the two worlds—the macrocosm and the microcosm. His great, unfinished work was titled *A History of the Two Worlds* (*fig. VII*).[20] The goal of the study of nature was the understanding of the greater and lesser worlds. In large measure, Fludd delved into natural philosophy because the "essence of God himself [is] masked from the sight of unworthy men with the material mantel of Nature."[21] An important key to understanding "Nature" was chemistry, because it reveals the components of the "material mantel" and exposes the processes that create it. Fludd

---

19 Joscelyn Godwin's brief biographical and intellectual study of Fludd is still the best overview available: Godwin, *Robert Fludd, Hermetic Philosopher and Surveyor of Two Worlds* (London: Thames and Hudson, 1979). Fludd, a Mosaical philosopher, was attacked, along with other Renaissance magi, by Marin Mersenne in his discussion of Genesis; see prologomena to *Quaestiones caleberrimae in Genesim* (Paris, 1623): "The attitude of Mersenne towards the new science is unique among the commentators.... He appeals for more knowledge of science by theologians" (Williams, *Common Expositor*, 179). Williams notes that Mersenne's commentary contains several scientific diagrams, although his purpose is theological.

20 Godwin, *Robert Fludd*, 9–11.

21 Allen G. Debus has written extensively on Fludd. He printed the manuscript of Fludd's philosophical key with a commentary. The quotation is taken from that edition. Allen G. Debus, *Robert Fludd and His Philosophical Key* (New York: Science History Publications, 1979), 88.

TRACTATUS SECUNDUS,
DE NATVRÆ SIMIA
ſeu Technica macrocoſmi hiſtoria,
in partes undecim diviſa.
AUTHORE
ROBERTO FLVDD ALIAS DE
Fluctibus, armigero & in Medicina
Doctore Oxonienſi.
Editio ſecunda.
FRANCOFVRTI,
Sumptibus hæredum JOHANNIS THEODORI
de BRY; Typis CASPARI RÖTELII.
ANNO M. DC. XXIV.

Figure VII. Title page. Robert Fludd, *Tractatus Secundus, De Naturae Simia* of *Utriusque Cosmi maioris salicet et minoris metaphysica.* Frankfurt, 1624.

believed that discovering the truths of nature is a profound act of worship. His philosophy was "Mosaicall" because it, like theology, revealed God. Correct knowledge of nature was knowledge of God. Only such knowledge could give man the power to act in the world in imitation of God. Fludd made this point "To the Reader" in the *Mosaicall Philosophy*:

> And yet nevertheless, lest mine intention should by the misprision of any, be ill interpreted or misunderstood, I think it convenient, to certify you, that my purpose, in the progress of this Sacred or Mosaicall Philosophy, is farre from any presumption, to trench upon, or derogate from the deep and mysticall Laws of Theology in her pure and simple essence, or to oppose the current of her Argument against those usuall Tenents and Authentick rules in Divinity, which have been long since decreed and ordained by the Ancient Fathers of the Church. But as it is certain, that one and the self-same place in Scriptures hath a two-fold meaning, to wit, an internall or spirituall, and an externall or literall; and either of these two senses are true and certain, though they seem to vary or differ by a diverse respect: no otherwise than under the name of one and the same man a double nature, namely a spirituall Soul, and a materiall Body, are really to be understood: So also besides such mysticall interpretations, as the Texts of Scripture do internally contain; it may also express and delineate externally such created realities, as belong unto the true Subject of the most essentiall Philosophy.[22]

Fludd distinguished his explication of the material world from "mystical interpretations." Although his thought may appear mystical to modern readers, he believed it to be solidly based upon experiment, or perhaps more correctly, upon "ocular" demonstrations.[23] By means of experiments correctly interpreted, he sought to overthrow "the falshood of the transitory and fading wisdom or Philosophy of the Ethnicks."[24] The true philosophy would be "grounded upon

---

22 Fludd, "To the Judicious Discreet Reader," *Mosaicall Philosophy* (London, 1659), unpaginated.

23 For an analysis of Fludd's use and understanding of experimentation, see D. Graham Burnett, "The Cosmogonic Experiments of Robert Fludd: A Translation with Introduction and Commentary," *Ambix* 46 (1999): 113–70.

24 Fludd, *Mosaicall Philosophy*, 1.

the eternall Spirit of Sapience."[25] By this Fludd meant an experimental interpretation of nature that was congruent with scriptural doctrines. He followed St. Paul in Colossians 2:8, arguing "that the Philosophy of men . . . is founded upon vain fallacy, as the traditions of Ethnicks, and according unto the elements of this world, and not according to Christ."[26]

Fludd began the *Mosaicall Philosophy* with a discussion of the weatherglass, an instrument to which he referred in several of his works. The glass, Fludd reported, was an ancient invention. It is made by inverting a heated, bulbed tube in a dish of water (*fig. VIII*). The water rises to a point leaving air trapped above it. The tube can be calibrated to mark changes in the weather as the water level rise or sink.[27] Fludd concluded from observation that the processes of heat and expansion that control the water level are fundamental principles of nature that were not understood by the Aristotelians. The Aristotelians had explained the change in water level in the weatherglass by attributing it to the attraction of water to the Sun's heat.[28] Fludd considered the action of the weatherglass to be ocular proof of natural principles that had been known to Moses and explained in Genesis, but that had been stolen and misinterpreted by the Ethnicks. According to Fludd, the correct understanding of the weatherglass originated in "Master Moses." Moses received his knowledge "by the finger of God." Fludd emphasizes that the glass was an ancient tool for philosophers:

> [Ethnick Philosophers did] assume and ascribe unto themselves those principles of their Philosophy, which of right did appertain unto the wise and divine Philosopher *Moses*, and did mask or gild over their theft, with new names or titles, which they imposed on them, the better to make a shew, that they were established by their own inventions, as shall be shewed hereafter. In like manner, the Instrument, commonly termed the *Calender*, or *Weather-Glasse*, hath many counterfeit Masters or Patrons, in this our age, who, because that they have a little altered the shape of the modell, do vainly

---

25 Ibid.

26 Ibid., 2.

27 Godwin, *Robert Fludd*, 60.

28 Ibid.

glory and give out, that it is a Master-piece of their own finding out. As for myself, I must acknowledge, and willingly ascribe unto each man his due, and therefore will not blush or be ashamed, to attribute justly my Philosophicall principles unto my Master *Moses,* who also received them, figured or framed out by the Finger of God. . . .[29]

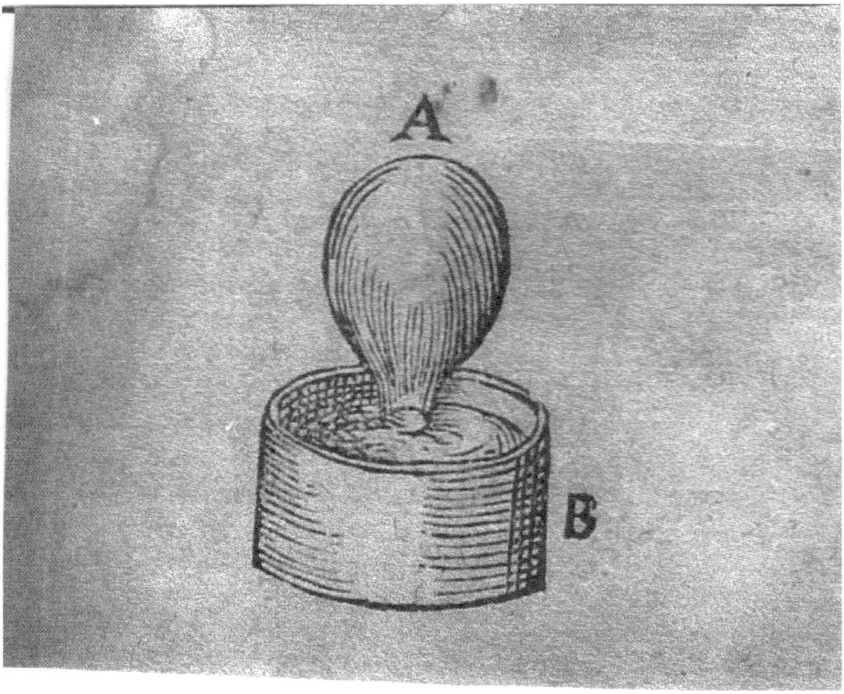

Figure VIII. Weatherglass. Detail from Robert Fludd, *Tractatus Secundus, De Naturae Simia* of *Utriusque Cosmi maioris salicet et minoris metaphysica.* Frankfurt, 1624.

---

29 Fludd, *Mosaicall Philosophy*, 3.

The weatherglass was a significant experimental-experiential instrument through which the natural philosopher observed the fundamental processes of nature. The air in the bulb tube behaved like the "universall sublunary element" air. The glass was a microcosm of the macrocosmic seasons, the weather in the world working in the same way, by the principles of expansion, contraction, sympathy, consanguinity, rarefaction, and condensation.[30] The principles are divine in origin and fundamental to the created world.

Ecclesiastes 1:4 teaches that God created the world by "Wisdom." Fludd's conception of wisdom as the agent of God is the same as that of Khunrath. Wisdom is the light of God and his light is diffused throughout the Creation. Indeed God's light is Jesus, who is the divine image by whom all things visible and invisible are made, as Paul stated in Col. 1:15–17:

> Christ is the image of the invisible God, the first begotten of every creature, because that in him, all things visible and invisible, in the heavens and in the earth, were made, whether they be thrones, or dominions, or principalities, or potestates, all were created by him and in him; and he is before all creatures, and all things consist in him.[31]

Jesus, as the Word, Wisdom, Light, the creative image of God, was the source of natural phenomena. Only by comprehending the source of Creation can the processes imbedded in nature be understood. Genesis is thus the point of departure in comprehending how nature works and in drawing back the material veil that conceals God from man. Fludd read Genesis with the eyes of a Christian chemical philosopher, for he believed that Genesis could not be properly comprehended without recourse to the Christian scriptures. These reveal that Jesus is the Logos and Light. In addition, only through understanding the role of Jesus in the world could a philosopher properly interpret the *Corpus Hermeticum* and the Christian Kabbalah. The finger of God taught Moses everything. What he learned was passed on to Joshua and the prophets and contained in the religious-mystical texts and the Kabbalistic tradition, which Fludd recognized as authentic.

Chemistry, broadly understood, was also a source of knowledge of the creative act. Fludd believed that Genesis teaches creation *ex nihilo*; but for him

---

30 Fludd used the term "*consanguinity*" to mean similarity of substances. Ibid., 7.

31 Quoted in Fludd, *Mosaicall Philosophy*, 14.

"nothing" implied a "certain matter in the highest mystery." "Before . . . that Spagiricall [chemical] separation, which by the Word of God, or divine Spirit, *Elohim* was effected in the six daies work of the Creation, . . . darkness was upon the face of the abyss."[32] The abyss was the substrate from which the material world was created. Fludd reports that the Kabbalists called the abyss *alephtenebrosum*, whereas Hermes said there was "an infinite shaddow upon the abyss, also water, and a thin intellectuall spirit were in the chaos."[33] Moses and Hermes, with their chemical creation from the chaotic waters, agreed with St. Peter, who wrote, "the heavens and the Earth were of the Waters and by the waters, by the Word of God."[34] To that of Moses, Hermes, and St. Peter, Fludd added the testimony of the Kabbalah as presented by Johannes Reuchlin. The Divine Spirit or Light which made all things proceeded from the "Ensoph or the fountain or root of infinity."[35]

The Creation described in Genesis was both spiritual and material. The spiritual Light chemically informed chaotic matter. The contradictory and complementary principles of hot-cold, expansion-contraction, and attraction-repulsion are spiritual and material. They are the fingerprints, so to speak, of God in nature, which can be seen in the weatherglass. Fludd believed that the complex nature of the Creation reveals aspects of the Creator that are found in scripture but are, perhaps, best articulated in the Kabbalah of Divine Names. The Divine Names, or Attributes, are found in the drawings of Kabbalistic trees, which are iconographic representations of the divine attributes and powers glimpsed at and in the Creation.[36] The First Name, according to Kabbalists, is *Ehieh*, Hebrew for "let there be." It is "the fountain or originall of all creatures." The Second Name is *Jehova*, the blessed and gracious emanation of the world, and the extraction of the humid or created nature, namely, the waters out of the womb or bowels of the dark and

---

32 Fludd, *Mosaicall Philosophy*, 44–46.

33 Ibid., 45. On page 47, Fludd has a long quotation from Johannes Reuchlin's *De Arte Cabalistica* (Hagenau, 1517) on creation and the dark aleph. This and other references to Reuchlin indicate the importance of *De Arte Cabalistica* in shaping Fludd's understanding of the Kabbalah as an ancient tradition.

34 Fludd, *Mosaicall Philosophy*, 67, citing 2 Peter 3:5.

35 Ibid., 171, citing Reuchlin, *De Arte Cabalistica*.

36 Fludd does not present an illustration of the tree in *Mosaicall Philosophy*, but gives several in *Utriusque Cosmi, Tomi secundi* (Oppenheim, 1619), 174 and 181.

deformed chaos. The Third Name is *Elohim*. This name is the source of prudence and it gave rise to the Sons of Elohim, which are not persons, but the divine instruments in the elemental world. By them the first creatures were made by means of the "contraries out of chaos," for it was by "the Spagericall or high Chymicall virtue of the word, and working of the Spirit, that the separation of one region from another . . . was effected or made."[37]

Fludd's success and rigor as a Mosaical philosopher was due in large part to his erudition. He used ancient texts to reveal a philosophy that was truly Mosaic. It was, perhaps, the most sophisticated and articulate presentation of a Christian chemical philosophy of nature in early modern Europe. Fludd took account of scripture, hermeticism, Kabbalah, and chemistry. Fludd, with his hermeneutic-experimentalism, differed from earlier chemists in that, by means of profound natural scientific, theological, and Kabbalistic learning, he better described how chemical processes are connected to the Creation. His interpretation of experiments is exhaustive and made in terms of his theory of nature and of his understanding of theology. He presents not-quite modern scientific, but rather ocular proofs of his understanding of nature. These proofs could be discussed, and falsified or verified. Fludd demonstrates the congruence of the various intellectual, hermetic traditions valued by Renaissance philosophers. They were congruent, in his mind, because they all originated with Moses, who was taught by the finger of God. From our historical perspective, their congruence arises out of their shared origins in the philosophical traditions of the ancient world.

III

Robert Fludd's interpretation of experiments in order to understand the Creation text of Genesis, while deeply imbedded in his catholic learning, did rely to a significant degree on quantification. Indeed, from Paracelsus on, chemists were increasingly aware of the importance of quantification. A study of the technical history of chemistry, which is not my interest here, reveals that quantification had been part of the art since its inception, although not rigorously practiced. Qualitative changes, such as changes in color, predominated in alchemy. The Paracelsian realization of the value of chemistry in

---

37 Fludd, *Mosaicall Philosophy*, 174–75.

understanding natural processes and in producing medicines was an impetus to greater technical proficiency. Rigor in the laboratory, which was required by the chemical philosophy, was ironically a factor in the refining and rejection of the many amorphous and unfocused doctrines of Paracelsus. The more specific and concrete concepts of matter and chemical processes in the seventeenth century are the result of increased emphasis on quantification. This did not mean, however, that the development of a Christian chemistry, begun by Paracelsus, was not also the goal of those who thought him old-fashioned, obscure, and just plain wrong. Such a "modern" chemist was Joan Baptista Van Helmont (1579–1644).[38]

Van Helmont was a physician who, not remarkably, saw chemistry as essential to creating healing medicines.[39] He was a technically adept and experimentally gifted chemist. He interpreted both traditional and his own original experiments in such a way that they yielded new and better understanding of chemical processes. Van Helmont was especially talented at quantification.[40] Van Helmont's chemistry was much more an outgrowth of

---

38 William R. Newman and Lawrence M. Principe in *Alchemy Tried in the Fire* (Chicago: University of Chicago Press, 2002) provide many insights into Van Helmont as a technical chemist and his influence on Starkey and Boyle. Their quotation from Lavoisier's *Essays Physical and Chemical* highlights Van Helmont's importance:

> We are astonished, in reading this Treatise [Joan Baptista Van Helmont's *Ortus medicinae*], to find an infinite number of facts, which we are accustomed to consider as more modern, and we cannot forebear to acknowledge, that Van Helmont has related, at that period, almost every thing, which we are now acquainted with, on this subject [i.e., "airs"]. . . . It is easy to see that almost all the discoveries of this kind, which we have usually attributed to Mr. Boyle, really belong to Van Helmont, and that the latter has even carried his theory much farther. (299)

My essay, "Boyle and Newton on the Transmutation of Water and Air, From the Root of Helmont's Tree," *Ambix* 27 (1980): 11–18, makes the same point. Allen G. Debus early on articulated the connection between Van Helmont and Boyle: Debus, "Fire Analysis and the Elements in the Sixteenth and Seventeenth Centuries," *Annals of Science* 23 (1967): 127–47.

39 Walter Pagel's *Joan Baptista Van Helmont* (Cambridge: Cambridge University Press, 1982) is the best biography of Van Helmont.

40 Newman and Principe, *Alchemy Tried in the Fire*, 58–91, discuss Van Helmont and quantification in detail.

his religious faith than vice versa. His intellectual biographer, Walter Pagel, presents him as a deeply religious man who felt himself called to develop a "Christian philosophy."[41] His originality and ability to quantify in chemistry were an outgrowth of his belief that everything was created, as the scriptures declare, by weight and measure (Wisdom of Solomon 11:17). Van Helmont was motivated by more than belief based on scripture. He had prayed to God for power to understand nature and had experienced dreams answering his questions about how to obtain knowledge.[42] Given this context, Van Helmont believed that knowledge gained from chemistry was both grounded in experience and received through revelation.

Genesis was central to the seminal-corpuscularian theory of matter that Van Helmont developed. Like his younger contemporary, Robert Boyle, Van Helmont was critical of all contemporary matter theory. He rejected the fundamental nature of the *tria prima* of Paracelsus and three of the four classical elements (retaining only *water* while rejecting air, earth, and fire). Van Helmont eventually decided that one "element," the primordial water of Genesis, is the only principle of composition, the source of the matter that formed all things. According to Van Helmont, each atom of water contains the three principles: salt, mercury, and sulfur. Each principle dominates in its quality depending on whether water was solid, liquid, or vaporous.[43] Common water, the water found in the created world, retains some of the nature of primordial water. In reading Van Helmont, it is difficult to determine if his study of Genesis gave rise to his idea of water as *the* element or if his work in the laboratory conditioned his reading of Genesis.

Van Helmont analyzed the elements both scripturally and chemically. He denied "that God created the four Elements," seeing "no where, anything

---

41 Pagel, *Joan Baptista Van Helmont*, 11.

42 Van Helmont, *Oriatrike* [English translation of *Ortus Medicinae*], trans. J[ohn] C[handler] (London, 1662), 14. *Ortus Medicinae* (Amsterdam, 1648) is essentially Van Helmont's collected works, published after his death by his son Francois Mercurius Van Helmont. The edition that I have used for the Latin text is the third edition published in Amsterdam in 1652. See *fig. IX* for title page.

43 Van Helmont admitted, therefore, that the *tria prima* might exist as principles of heterogeneity, i.e., forms of basic matter that account for the diverse appearance of things, but rejected them as principles of composition. See Newman and Principe, *Alchemy Tried in the Fire*, 64.

is read of the Creation of fire."[44] Fire, as observed in the laboratory, does not "materially concurre into the mixture of bodies." Earth, another purported element, is actually a mixture of sand, minerals, and metals.[45] Air and water are not separate elements in Genesis; the text uses the word *shamyim* (heavens) for both the waters above (i.e., air) and the waters below (i.e., terrestrial water).[46] This being the case, water alone is the substrate or principle of composition according to the scripture. Van Helmont verified this by means of his famous tree experiment (discussed below).

Van Helmont's religiosity predisposed him to use Genesis to interpret his laboratory work. His theory, that seeds act on inert matter in the presence of ferments, enabled him to explain various chemical reactions and to recognize the existence of gas—one of his most important discoveries. One experiment produced an invisible explosive substance from coal. He knew that this substance was not a vapor, such as that produced by boiling water, but something unrecognized by other chemists. He called it *Gas*. The discovery of gas gave Van Helmont an insight into Genesis and the function of primordial Water:

> Gas and Blas are new terms introduced by me which were unknown to the ancients [the German translation defines *Gas* as a subtle water vapor or spirit and *Blas* as a wind or movement from heavenly bodies]. Gas and Blas are among the first physical necessities; therefore, these new concepts must be amplified and expanded. First, it must be shown how Gas is produced from water, that it is different from [the products produced by] heating water until vapor rises. To understand how this happens, one must learn the anatomy of water; therefore I will repeat that the most glorious God in the beginning created the heavens and the earth and the abyss of waters. The abyss began at the vault of heavens [the Dutch reads "at the far end of the dwelling place of the souls"] and ends above the globe of the earth. Nothing is written about the creation of air, which is also a body and was made into an element. Only after the six days

44 Van Helmont, *Oriatrike*, 48.

45 Ibid., 56.

46 Ibid., 48.

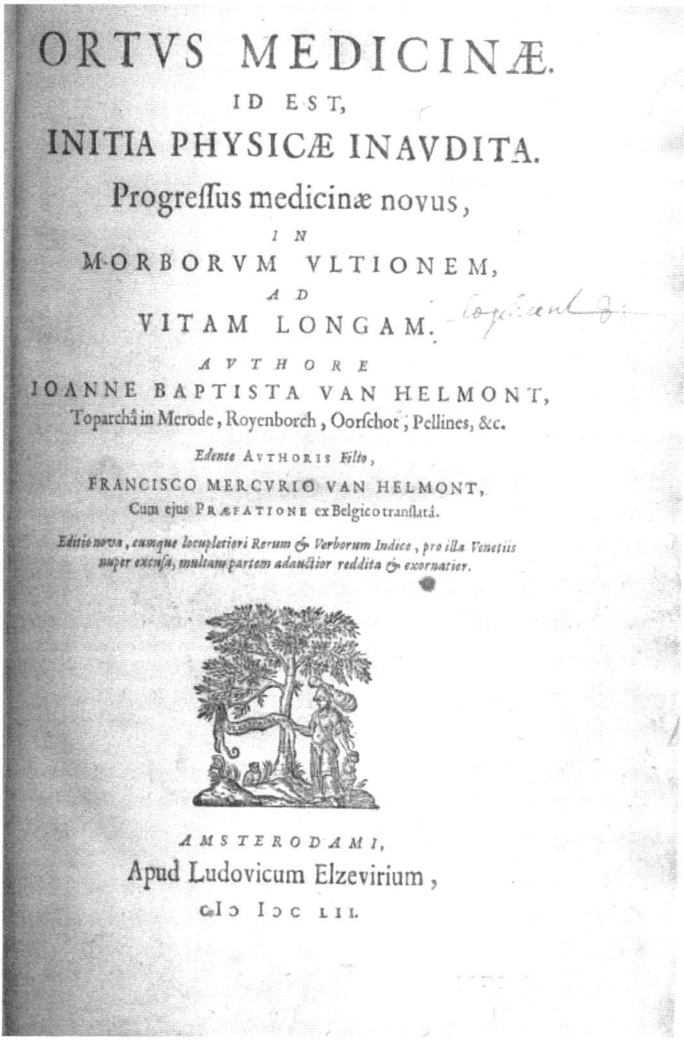

ORTVS MEDICINÆ.

ID EST,

INITIA PHYSICÆ INAVDITA.

Progressus medicinæ novus,

IN

MORBORVM VLTIONEM,

AD

VITAM LONGAM.

AVTHORE

IOANNE BAPTISTA VAN HELMONT,

Toparchâ in Merode, Royenborch, Oorschot, Pellines, &c.

Edente AVTHORIS *Filio*,

FRANCISCO MERCVRIO VAN HELMONT,

Cum ejus PRÆFATIONE ex Belgico translatâ.

Editio nova, cumque locupletiori Rerum & Verborum Indice, pro illa Venetiis nuper excusa, multam partem adauctior reddita & exornatior.

AMSTERODAMI,

Apud Ludovicum Elzevirium,

CIↃ IↃC LII.

Figure IX. Title page. Joan Baptista Van Helmont, *Ortus Medicinae*, 3rd ed. Amsterdam, 1652.

> of creation, and there was a place [for it] to fill, was there air. Thus heaven means air and the matter of the heavens is air, something not known before [I discovered it]. [The Dutch reads "and after the beginning, the air was with the water together in one body and known by one name, heaven"]. The Eternal created the eternal firmament and separated the water which ought to remain below from the water which ought to remain above. Now this firmament was not a cataract or passive division of the waters, but by the power of its operation, the very principle of separation. Just as the sun is not a dividing wall between day and night but is rather itself the source of division, the heavens or air is the source of division between the waters.[47]

Air and water, as the Hebrew text of Genesis proves, are the same thing, one element and the basis of all material things. Among the things formed from water-air, Van Helmont discovered, are invisible material bodies, namely, gases. The gas of water is particularly relevant to his discovery of gases because Genesis implies its existence:

> Now I will turn to the history of *Halitus* under which term both steam and Gas can be meant. It is nothing more than the examination of what is in air. By Gas, I do not understand the dry, oily body the ancients called exhalation but rather another watery, vaporlike [body] which in the past was associated with the substance of meteors. I consider [Gas] the body [corpus] of water, containing the elements in it and the spirit of Mercury, a liquid, most simple, a tasteless, simple salt of water. It contains in itself a uniform, homogenous, simple and inseparable sulphur.[48]

Van Helmont believed that the discovery that air and water are alike, for they are the substrate from which all things were originally created and from which all things are still formed, was fundamental to understanding chemistry. The principles of composition contained in air and water continued to renew the world. Chemistry studies and uses the creative

---

47 Van Helmont, *Aufgang der Artzney-Kuenst* (Sulzbach, 1683), 109. The *Aufgang* is an interpretive translation of the *Ortus*. The passage in the *Ortus* begins on page 50. The translation is mine, based on both texts.

48 Van Helmont, *Ortus*, 60; *Aufgang*, 110.

combinatory powers that were implanted in the world when the spirit of God moved on the face of the waters. Van Helmont's discovery of subtle vapors or gases grew in part out of his reading of Genesis. His chemical experiments were at one level a commentary on Genesis.

Van Helmont's famous tree experiment proved that water retained its creative potential. For five years, he grew a tree that eventually weighed 169 pounds. Before planting the tree, Van Helmont weighed the soil. He found that the soil in which the tree grew diminished by only two ounces from its original weight. Quantitative analysis convinced Van Helmont that the watering of the tree had supplied the greater part of its growth. The experiment also indicated, as did the growing of plants in water alone, that seeds, often themselves invisible, were vital principles necessary to transforming inert matter They were that which formed the water and air into concrete things. The chemist needed to understand the interaction of water and seed in order to imitate nature. Van Helmont connected his coal-gas and tree experiments thus:

> Therefore, coal gas, which I do not dare light but in an open vessel, likewise with its burning makes pure water. Seeds, indeed, are the property of concretion which abide in Gas . . . when it returns to [its] original water. That all plants arise immediately and substantially from the element of water alone I learned from the [tree] experiment.[49]

Van Helmont's experiments were carefully constructed and carried out. He paid careful attention to quantity, weight, and measure. He usedprimitive thermometers to measure changes in temperature (*fig. X*). His results were replicable by others. His interpretation was suggested by the results. He discovered new things, such as gases. Van Helmont rejected the theories of both the ancients and Paracelsus in an attempt to introduce greater rigor and accuracy to chemistry. All of this, however, was in the service of creating a "Christian philosophy." Van Helmont was devoted no less than Fludd to a proper understanding of nature through experiment and scripture. Genesis was an essential text for Helmontian chemistry. He saw it as the source of his matter theory and his practical discoveries. His chemistry, no matter how practical, was theological. His laboratory was figuratively enveloped in sacred

[49] Van Helmont, *Ortus*, 88.

52 A

tra meum libellum de Spadanis, impertinenter
uſus eſſet meo inſtrumento : attamen non deſiit
me ſuis ſtupiditatibus immiſcere. Dicit nam-
que. Me voluiſſe edere motum perpetuum.
Nec cognovit quidem in eo, ſe ſibi contradi-
cere. Nam in inſtrumento ceſſat motus, poſt-
quam aqua adſcenderit, aut deſcenderit juxta
temperamentum ambientis. Neque enim plus
iſte motus dici poteſt perpetuus, quam velum
templi, ventorum mutationi deſtinatum. De-
tegit ergo Heer, ſe neſcire, quid ſit motus per-
petuus. Divulgaveram enim meum organum,

ut juxta doctrinam meam de fon-
tibus Spadanis probarem, quod aër
ſuſtineret vulgatam rarefactionem,
& compreſſionem citra mutatio-
32 nem ſui elementi. Dicit enim
Heer : Spiritum vitrioli, penetraſſe
ſubſtantiam vitri; quod nemo con-
cedet, qui longe ſubtiliores liquo-
res norunt ſervari, etiam in igne.
Nec juvit, quod ipſi affirmaveram,
quod Clavius in Collegio Societa-
tis Roman. ante annos 60 aquam in
vitro recluſerat, hujus figuræ, cu-
jus ne minima gutta perierat. Id
quod linea, per adamantem ibidem ſignata,
oſtendit. Heer autem apud Idiotas oſtentabat,
ſe quandoque Patavii fuiſſe Matheſeos Profeſſo-

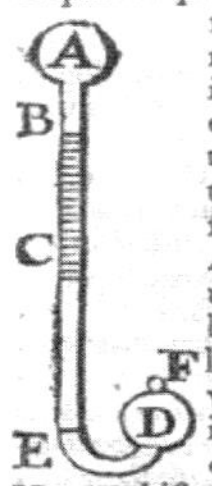

rem. Quare volui in charta de-
monſtrare, ipſius omnimodam
ignorantiam Matheſeos. Sed factus
eſt coram mutus. Ad demonſtra-
tionem verò ſic proceſſi. Delinea-
ta enim figura vitri, in quo A & D
ſunt duæ ſphæræ, repletæ aëre.
A autem & ſuperior, eſt exterius
undequaque clauſa. D verò eſt glo-
bus inferior, apertus in fine Cana-
lis F. Sunt autem A & D ex unico
vitro connexa per canalem B C E.
in quo eſt liquor vitrioli, ne fraceſ-
ceret, quem dixeram percontanti
Heer, rubificatum maceratione roſarum, ut vi-
ſibilior eſſet. Rubens ergo ille liquor, eſto B C.
Super hanc ergo figuram conſtitui hoc pro-
blema.

Liquor B C non poteſt ſe movere per tem-
peramentum ambientis, in canali, niſi unus glo-
borum ſit apertus, alter verò clauſus. Expli-
catio dati. Sit vitrum cum liquore, ut ſupra de-
ſcriptum. Explicatio requiſiti. Oſtendendum.
Quod liquor B C non poſſit moveri in canali

Figure X. Thermometer. Detail from Joan Baptista Van Helmont, *Ortus Medicinae*, 3rd ed. Amsterdam, 1652.

writ. His mind sought a chemistry that elucidated Genesis. Only such a chemistry could be true and produce healing medications.

IV

Van Helmont connected his experimentalism to nonchemical texts. Those texts were religious. He prayed, dreamed, and worked in his laboratory. Johann Rudolph Glauber (1603/5–1666/70) was less overtly spiritual than was Van Helmont. He was a self-educated, and preeminently practical, chemist. He was also a proto-industrial chemist. His writings deal with applied chemistry, but they also express then-current chemical theories. Nonetheless, he was deeply religious and accepted Genesis as important to chemical theory. Glauber demonstrates how rooted Genesis had become in chemical thought. The biblical creation account performed several functions for Glauber. First, it established chemistry as a Christian discipline. Second, it inspired matter theory. Third, it set the pattern and provided limitations on ideas concerning elements and creative processes.

Glauber did not believe that art (i.e., alchemy) could "create Gold or Silver, but Nature can."[50] In other words, chemistry is limited and can only imitate the stronger forces built into nature. God's creative act(s) remain part of nature and function more potently than human actions. He describes the creation of metals as having been begun by God:

> For the Metals are not all together created in the beginning of things, but begotten in length of time, out of the bosome of the Elements; and on them, being created by the Omnipotent GOD, is this Command Injoyned, and this Power implanted, that they should give growth to all things, by their Vertue and Efficacy; for accomplishing of which thing, the one cannot in the least want the Company of the other.[51]

For example, the element of fire develops metallic seeds in water.[52] Water is the source of metallic materiality. That materiality is achieved by the action of

---

50 Johann Rudolph Glauber, *The Mineral Work*, in *The Works*, trans. Christopher Packe (London, 1689), 121.

51 Glauber, *Birth and Anatomy of Metals*, in *The Works*,115.

52 Note the intimation of a Khunrathian idea of fire.

seeds that were made at the Creation. The action of seeds is a stable process that will endure until the end of the created world:

> For the Stars or Elements of Fire, delivers out the metalline Seed out of its own bowels; which the air carries down into the Water, that it may adapt to it self, a palpable form or body, which the Earth (embraceing it) doth cherish, nourish, and augment from form to form, until it comes to be a perfect Metal, which it (at length) brings forth into the light, as a Mother doth her mature young one; which Conception and Generation of the Metals, taking its Original at the very beginning of the World, will alwaies continue even unto its Dissolution.[53]

Glauber's recognition that metals persist (are elemental?) came both from his understanding of Genesis and from his laboratory work. Genesis teaches that God made the world and everything in it from chaos. Each created thing has its place. The world, by God's design, became self-sustaining, exhibiting a "perpetual circulation" that is seen in the processes discovered by natural philosophy (or chemistry):

> Thus therefore stands the case: We read in *Moses*, in the first of *Genesis*, that God, when he made the World out of the confused *Chaos*, did give the Elements their original first, and assigned to Earth its proper place, and injoyned on earth its Office to be done; but by what means they are preserved by the interceding perpetual Circulation, natural Philosophy doth demonstrate.[54]

Moreover, religion dictates that the chemist use chemistry for the material well being of others. Glauber cites the golden rule before his discussion about how grain extracts could be used to nourish sailors on long voyages. He composed his famous treatise, *The Prosperity of Germany*, in order to show how chemistry could result in the better use of wine, grain, and wood. He began that work with a theological statement that nature reveals God's will to men for their betterment.[55] The study of nature is not only theologically valuable for learning about God's will, but also leads to material benefits that strengthen spirituality:

---

53 Glauber, *Birth and Anatomy of Metals*, in *The Works*, 115.

54 Ibid., 119.

55 Glauber, *Prosperity of Germany*, in *The Works*, 294–440.

> The Most Excellent Glorious God, and our most Merciful and Bountiful Father hath (out of his great Love unto, and Fatherly Care of us, from Eternity) set Nature before our Eyes, that he might signifie to us his Will; which said Nature, is no other thing but the proper, eternal, and only Will of God. To this Nature, God hath (as to his obedient and faithful Instrument) absolutely delivered, and committed his Will and Command to be by her faithfully and diligently performed; she accomplisheth the Task imposed on her, without intermission, nor resteth she one moment of time from doing the Divine Will with her utmost study and endeavour.[56]

Nature was created to serve man, so that man would serve God:

> 'Tis therefore clearly and evidently manifest, that God Created all things for the service of man, that he might thereby acknowledge the Divine Bounty, and Honour, and fear the Creator, and bestower of all good things, and yield him due Thanks for his Gifts.[57]

Glauber expressed his most thoroughgoing theory of chemistry in this work on Germany's welfare. "The matrix [note the Paracelsianism] of all things is water," he stated. The chemical artisan must use male-like *astrums* and feminine *astrums* and *matrices* from which the world was created. The chemist needs to imitate the creation or creative forces of the great world in the lesser world of his chemical art:

> After this manner the World itself was Created, where the Spirit of God moved upon the Water and impregnated it; and even as we daily see that the stars do impart to the inferiour things their virtues, and they are thereby multiplied; even so may we do the same by art in these inferiour things; for such as are above, are also beneath. Those above are beyond our reach; but the inferiour ones, no so: For art can do the same with or upon the inferiour things, as the Stars upon the superiour ones.[58]

---

56 Ibid., 294.

57 Ibid.

58 Ibid., 363.

The relationship between the world above and the one below, which Glauber read into Genesis, was explicit in traditional alchemy. The *Emerald Table* (composed by Hermes according to Glauber), Paracelsus, and Basil Valentine all testified to the influence of the superior on the inferior:

> As *Hermes* doth clearly demonstrate in the *Smaragdine* [Emerald] Table. He that understands these my sayings, will also understand those things, which *Paracelsus* delivers in his Book of the *Vexation of Alchymists*, in the 5th Rule especially of ☽; and *Basil (Valentine)* where he treats of Sulphur.[59]

Glauber lived and wrote a century after Paracelsus. He was heir to practical and theoretical developments in chemistry; yet he, like Paracelsus, was committed to a Christian natural philosophy. Glauber wove into his works Genesis, along with the doctrines of Hermes, Paracelsus, and subsequent chemists. The seamless connecting of scripture with the "art" was a commonplace in the seventeenth century. Glauber linked Holy Writ to his chemistry not as an affectation, but as a perceived necessity. Genesis was essential to his theory of matter and his search for the fundamental processes by which a chemist could use nature to benefit mankind.

Khunrath, Fludd, Van Helmont, and Glauber differed greatly from early Paracelsianism and from each other. Whether in theory or in practice, however, they shared the Paracelsian quest to articulate the Mosaical philosophy and to supplant pagan thought with a true Christian understanding of nature. Jesus, as the light of the world, offers the only certain hope to comprehend natural processes and to use them to practice the golden rule on a grand scale. The search for the Mosaical philosophy began in scripture and moved to the study of nature. Nature is the garment that has clothed the divine since the beginning. The philosopher could remove nature's garment and reveal the divine using knowledge gained through a new epistemology, one based upon spiritual insights. From that epistemology, methods could be derived that would reveal natural processes. The new epistemology used by virtually all of the sixteenth and seventeenth century chemical philosophers discussed above is the subject of the following chapter.

---

59 Ibid.

5. Soul Knowledge: An Epistemology of Illumination

The prophetic tradition informed the Christian search for spiritual truth. Peter, whose knowledge that Jesus was the Son of God was not revealed by "flesh and blood" (Matt. 16:16–18) and Paul's vision of Christ on the road to Damascus (Acts 9:1–6) established mystical-revelatory experiences as a source of Christian truth. The writings of Pseudo-Dionysius and the medieval prophecies of Hildegard of Bingen and Joachim of Fiore were widely studied as guides by those who wished to penetrate the depths of the scriptures. Thomas à Kempis's *Imitatio Christi* was a manual for inner illumination.[1]

Renaissance and Reformation intellectuals and spiritual leaders were imbued with the spiritual illumination tradition both from the Middle Ages and from the Neoplatonism that was reemphasized in the early fifteenth century, beginning with the visit of Plethon to Italy. Neoplatonic mysticism, especially orphic chanting, was a path to deeper knowledge. Although very different individuals, Ficino, Pico, and Luther were each seeking knowledge

1 The importance of medieval "prophets" in the Renaissance and early modern period has a large literature. An enduring classic is Norman Cohn's *The Pursuit of the Millennium* (New York: Essential Books, 1957). A thoughtful essay on the subject germane to this study is Christopher Bamford's "The Meaning of the Rose Cross," in *The Rosicrucian Enlightenment Revisited*, ed. Ralph White (Hudson, NY: Lindisfarne Books, 1999), 43–72.

that went beyond that gained from a superficial reading of scripture.[2] The idea of a profound, personal revelation as a source of divine insight perhaps reached its height in the Anabaptists, who were seen as leaping beyond the boundaries of theology and tradition in an heretical quest for inner light.[3]

The chemical philosophers, Paracelsus, his disciples, and critical successors rejected the pagan, "ethnick," study of nature found in the universities. They favored a return to the Holy Scriptures as a guide to the study of nature. Chemical philosophers believed that they could comprehend nature only by reliance on the light of nature or divine illumination to interpret their personal experiences of nature. This epistemology of illumination was drawn from the Christian-Neoplatonic tradition. Its goal was to produce a correct reading of the book of nature, just as theologians used it to penetrate the book of scripture.

I

In light of the success of mathematical astronomy and physics at the hands of university-trained natural philosophers such as Puerbach, Copernicus, Kepler, and Galileo, the chemists' complaint about the Aristotelian tradition and the universities may seem unfounded, if not detached from reality. The chemists, although committed to weight and measure as mandated in scripture, did not believe celestial-geometrical abstractions were useful to their art. They were Christian philosophers who sought experience of the divine in nature and who worked in the laboratory with fire, trying, by art, to imitate the creation. Experience and observation, not abstractions, were their sources of data.[4]

---

2 Walker's *Spiritual and Demonic Magic* covers the Neoplatonic revival in detail. Joscelyn Godwin discusses the integration of pagan and Christian thought in *The Pagan Dream of the Renaissance* (Grand Rapids, MI: Phanes Press, 2002).

3 The development of early Anabaptism with its spiritualism and Lutheran reaction to it is the subject of Hans J. Hillerbrand, ed., *Radical Tendencies in the Reformation* (Kirksville, MO: Sixteenth Century Journal Publishers, 1988).

4 Above I have argued that the chemical philosophers' experiencing nature resulted in a development of the idea of experiment. Experiment or experience was not a new concept, but its use was greatly intensified in the sixteenth and seventeenth centuries. William R. Newman, *Promethean Ambitions: Alchemy and the Quest to Perfect Nature* (Chicago: University of Chicago Press, 2004), argues that Aristotelian science was experimental. This is a valuable insight. Nonetheless, chemical experimentation increased in the

Interpretation came from the spirit or, more precisely stated, through the understanding of the soul, which had been created in God's image. Although most of the chemists were university trained, like religious reformers, they rejected university scholasticism in favor the Christian revelatory tradition. What were the Aristotelian epistemology and physical theories that they so consciously and conspicuously rejected?

Until the mid-seventeenth century, Aristotelian physics and matter theory dominated the study of natural philosophy at European universities. The four elements, earth, air, fire, and water, in varying proportions, accounted for all things in the sublunar world. The celestial realm eternal in nature and exhibiting uniform circular motion, was made of a fifth element, the quintessence. Things fell down because the Earth was the center of the world.

---

sixteenth and seventeenth centuries among chemical philosophers and among their opponents, such as Libavius. The epistemology discussed here was an important factor in that phenomenon. But it was not the only factor, as the case of Libavius demonstrates. Libavius's work was greatly influenced by his polemics against the chemical philosophers and their theological-spiritual approach to the study of nature.

The writings of Francis Bacon, especially *Sylva Silvarum or a Natural History in Ten Centuries* (London, 1627), influenced the methodology of natural philosophy and contributed to the formation of the Royal Philosophical Society. Bacon partook of the critical attitude to Aristotelianism and promoted the idea of a Christian society of virtuosi. Frances Yates, in her speculative and controversial *The Rosicrucian Enlightenment,* (158–69), described Salomon's House in the *New Atlantis* (1626) as related to the Paracelsian-influenced Rosicrucians and their utopian scientific society. Paolo Rossi has presented a more-detailed discussion of Bacon and his program for the development of natural philosophy in *Francis Bacon: From Magic to Science* (Chicago: University of Chicago Press, 1968) [Italian original, 1957]. Stephen A. McKnight, who reminds us of how deeply imbedded Bacon was in the late sixteenth- and early seventeenth-century historical and religious context, argues against those scholars who overemphasize Bacon's modernity in *The Religious Foundations of Francis Bacon's Thought* (Columbia, MO: University of Missouri Press, 2006). As a critic of Paracelsianism and Neoplatonism, and as a nonchemist, Bacon falls outside this study, but, as I have shown elsewhere, he is among those who accept the concept of divine illumination as an agent in the restoration of a correct methodology for natural philosophy. See Michael T. Walton and Phyllis J. Walton, "Being Up Front: The Frontispiece and the *Prisca* Tradition," *Cauda Pavonis: Studies in Hermeticism*, n.s., 17.1–2 (1998), 8–13. John Henry, *Knowledge Is Power: How Magic, the Government and an Apocalyptic Vision Inspired Francis Bacon to Create Modern Science* (Cambridge, Eng.: Icon Books, 2002) is an important contribution to understanding Bacon in the context of early modern natural philosophy.

Any nonnatural movement (i.e., not down) required the application of force. As soon as the force impressed on an object was exhausted, it fell. Air and fire, by their nature, moved upward. Living organisms were composed of matter (the four elements) and indwelling directive cause, or telos. The telos was responsible for the particular form of an organism. Natural phenomena were analyzed using first principles, such as circular celestial motion, natural motion toward Earth's center, and the combination of elements. Mathematics, especially geometry, and the syllogism were primary analytical tools. A. C. Crombie outlined the Aristotelian approach to nature in his study of Robert Grosseteste.[5] In his commentary on Aristotle's physics, Grosseteste wrote:

---

5 A. C. Crombie, *Robert Grosseteste and the Origins of Experimental Science, 1100–1700* (Oxford: Clarendon Press, 1953). Crombie's *Medieval and Early Modern Science*, 2 vols. (New York: Doubleday Anchor Books, 1959), traces the development of methodology and epistemology through the scientific revolution. Rather than seeing the role of illumination in chemical epistemology, he believed that late medieval theologians "liberated rational and empirical inquiry into the laws that nature in fact exhibits" (2:315). This inquiry freed science from theology and led to a de-emphasis of causation and an emphasis on experiment in natural philosophies. He used Boyle as an example of this trend (2:317). I believe, in light of this study and recent Boyle scholarship, that Boyle's epistemology is related to that of the chemical philosophers, especially Van Helmont, more than it is related to medieval philosophy and mathematical astronomy. See Michael T. Walton, "Boyle and Newton on the Transmutation of Water and Air," *Ambix* 27 (1979): 145–69, also found in *Alchemy and Early Modern Chemistry, Papers from "Ambix,"* ed. Allen G. Debus, (London: Society for the History of Alchemy and Chemistry, 2004), 477–84; Lawrence M. Principe, *The Aspiring Adept: Robert Boyle and His Alchemical Quest* (Princeton, NJ: Princeton University Press, 1998); Michael Hunter, *Robert Boyle, 1627–91: Scrupulosity and Science* (Woodbridge, Suffolk: Boydell Press, 2000); and Newman and Principe, *Alchemy Tried in the Fire.*

Johannes Kepler's discoveries were motivated by a revelatory experience of the cosmic harmony. He differed from Fludd in that he believed that mathematics had to be tied to observation rather than prevail on a theoretical basis. See Michael T. Walton and Phyllis J. Walton, "The Geometrical Kabbalahs of John Dee and Johannes Kepler," in *Experiencing Nature*, ed. P. H. Theerman and K. Hunger Parshall (Dordrecht: Kluwer, 1997), 43–59. Irving Kelter argues persuasively that Tommaso Campanella and Federico Cesi regarded Copernicus and his supporters, including Galileo, as "Mosaical" reformers. See his two unpublished papers presented at the Sixteenth Century Studies conferences: "Pythagoras was a Jew? Renaissance Perceptions of the Ancestry of the Heliocentric Cosmology," October 1996, and "Moses as Cosmologist; the Use of Genesis as a Book of Science," October 1995.

> Since we search for knowledge and understanding by means of principles, in order that we may know and understand natural things we must first determine the principles pertaining to all things. Now the natural way for us to reach knowledge of principles is to start from universal applications and go to these principles, to start from wholes corresponding to these very principles.... Every agent has that which is to be produced, in some way already described and formed within it, and so the "nature" as an agent has the natural things which are to be produced in some way described and formed within itself. This description and form (*descriptio et formatio*), existing in the nature itself of the things to be produced, before they are produced, is therefore called knowledge of this nature.[6]

Sixteenth-century Aristotelianism was not appreciably different from that of Grosseteste. The noted Jesuit scholar, Christoph Clavius, rejected the heliocentric theory on the basis of Aristotelian first principles. He attacked "the absurdities that follow from [Copernicus's] position."[7]

Copernicus's absurdities were positing that the Earth moves and that the Sun stands at the center of the world. These conclusions contradicted philosophy and observation. If Earth has a daily rotation, why do objects that are dropped not fall to the east? If Earth is not at the center, why do things fall down? Clavius invoked such arguments to defend Aristotelian first principles and observation in a rigorously logical attack on heliocentricity.

In chemistry, the Aristotelian elements caused material qualities. For example, something burned because it contained the element of fire. Alchemy combined the four elements with its spiritual male and female forces. The medieval pseudo-Geber, noted above, and the chemical philosophers doubted that the four elements could account for their observations. Many chemical philosophers were also uneasy that Aristotle taught that the world was

---

6 Crombie, *Robert Grossetestei*, 55. Crombie also uses this quotation in *Medieval and Early Modern Science*, 2:12–13.

7 Christoph Clavius, *In Sphaeram Ioannis de Sacro Bosco Commentarius* (Venice, 1596). Translation mine. Williams, in *Common Expositor* notes that Hieronymus Zenchius, in *De Operibus* (Neustadt, 1591), interprets Genesis in terms of Aristotelian physics and may have consulted Clavius (178). He also states that Benedictus Pererius, in *Commentariorum . . . in genesem* (Rome, 1601), refers to Clavius on Sacro Bosco (186).

uncreated. Although this vestige of paganism had long been rejected by the church, it tainted Aristotle's entire philosophy.

Chemical philosophers found Aristotelian formalism inadequate to describe the real world. Syllogisms, geometry, and incorrect first principles yield nothing. Only a Christian philosophy inspired by God could produce a proper understanding of the world. The image of God is present in nature, which is his creation, and open for all to see. Paracelsus declares in his *Hermetic Astronomy* that, as God revealed truth to the apostles and prophets, so he also bestows such gifts "on astronomers and physicians."[8] Causality and logic are sterile compared to divine wisdom. Oswald Croll, mirroring his master, stated that "Scholastic Doctors, . . . for want of wit could never find the way into Nature's Sanctuary."[9] According to Croll, the way into the temple of the creation is through chemistry, a divine art that was condemned by scholastics:

> Let no man wonder if these insolent and open contemners of Nature's secrets [the scholastics], who tremble at the very name of *Chymistry*, are not afraid hitherto with a proud disdain, shamelessly to scoff at this Divine Art, slighting it with a sottish and barbarous kind of arrogance. . . .[10]

Scholastics could not penetrate nature because they did not understand how to know and by what means knowledge is obtained. A true physician or student of nature "should be born out of the Light of Grace and Nature of the inward and invisible Man, the internall Angell, the Light of Nature."[11] One prepared for this insight through chemistry. Experience or observation is transformed into knowledge or understanding of nature by the "internall Angell," which is the indwelling light of nature of the soul.

---

8 Paracelsus, *Hermetic and Alchemical Writings of Paracelsus*, 2:287.

9 Quoted in Pinnell, *Philosophy Reformed*,15. Pinnell translated Croll's preface to his *Basilica Chymica* and Paracelsus's *Philosophy to the Athenians*. These texts present the chemical philosophy and its fundamental principles clearly. Pinnell, therefore, could have a reasonable hope that philosophy would be reformed and improved in England as a result of his efforts. Owen Hannaway, in *The Chemists and the Word: The Didactic Origins of Chemistry* (Baltimore: Johns Hopkins University Press, 1975), discusses the epistemological dimensions of early modern chemical thought.

10 Quoted in Pinnell, *Philosophy Reformed*, 14.

11 Ibid., 22.

Peter Severinus, the famous Danish physician, in his explication of Paracelsianism, describes how the chemical physician should gain experience of the created world:

> [S]ell your lands, your houses, your clothes and your jewelry; burn up your books. On the other hand, buy yourselves stout shoes, travel to the mountains, search the valleys, the deserts, the shores of the sea, and the deepest depressions of the earth; note with care the distinctions between animals, the differences of plants, the various kinds of minerals, the properties and modes of origin of everything that exists. Be not ashamed to study diligently the astronomy and terrestrial philosophy of the peasantry. Lastly, purchase coal, build furnaces, watch and operate with the fire without wearying. In this way and no other, you will arrive at a knowledge of them and their properties.[12]

John French, the translator into English of the works of the Paracelsian Micheel Sandivogius, like Paracelsus and Croll, looked to the divine as the interpreter of the data gathered from the kind of observations Severinus prescribed. In his preface to Sandivogius's *New Light of Alchemy*, he states that alchemy "is the gift of God, and truly is not to be attained to, but by the alone favour of God, enlightening the understanding."[13] The Reverend Thomas Tymme, in his preface to his work on Quercetanus's chemical medicine, states clearly why alchemy is a gift from God. All knowledge has its origin in Genesis and the fire of creation. The alchemist-chemist work to understand God and His creation. The "ethnicks" deal only with their "reasons." Moses is the correct guide to knowledge, if we understand his writings. Even though Moses does not mention fire specifically, it is implied in the nature of the heavens. The ethnicks miss this crucial knowledge, because they are taken up with discovering causality:

---

12 Peter Severinus, *Idea medicinae philosophicae*, 3rd ed. (Hagae Comitis, 1660), 39, quoted in Allen G. Debus, *The English Paracelsians* (New York: Franklin Watts, 1966), 20. Debus's work revolutionized the study of early modern chemistry and touches upon the themes in this essay in many ways. What is presented here and in previous chapters is a detailing of his groundbreaking efforts and would have been impossible but for his many works.

13 F[rench], preface to his translation of Sandivogius's *New Light of Alchemy*, A4 [iii]. This work is discussed in chapter 3.

> But forsomuch as Moses in the first Chapt. of his *Genesis* (wherein he sheweth the creation of all things) maketh no mention of fier: it is more convenient that we leave it rather to the opinion of divine Prophet, then to the reasons of an *Ethnick* Philosopher. And therefore wæ acknowledge no other fier then Heaven, & the fiery Region which is so called of burning.[14]

Thus, a proper reading of Genesis demonstrates that the heavens are fire. Moreover, the use of fire in the laboratory is a key to understanding the physical creation.

II

The belief that illumination and experience are the keys to understanding and controlling nature is characteristic of chemical texts, but they rarely provide information on the process of gaining illumination. The works of three philosophers, Valentin Weigel, Joan Baptista Van Helmont, and Johann Joachim Becher, are exceptional in that they describe the specifics of illumination. Weigel (1553–88) was a German Lutheran minister influenced by Paracelsus and occultism. His nonconformist leanings were discovered after his death, when all of his works were published. Declared heretical, his works were burned in 1624. Heretic or not, Weigel's epistemology was rooted in Christian scripture and a chemical reading of the book of nature. The basics of his worldview were popularized in England with the publication of *Astrology Theologised* in 1649. Weigel is unlikely to have written the original of this work, but it presents his essential thought in much the same way as *Philosophy to the Athenians* presents Paracelsianism.

Weigel's theology of knowledge grew out of his understanding of God's nature. The macrocosm was a reflection of its soul, the light of nature, and man's soul as the habitation of the Divine Spirit—a microcosm of the great world. God is defined as containing "all things in himself."[15] He, however, is

---

14 Quercetanus, *The Practice of Chemicall and Hermeticall Physicke*, chap. 10, G3. How this text is the work of Tymme and how it diverges from Quercetanus is noted in chapter 3.

15 Valentin Weigel, *Astrology Theologised* (London, 1649); repr., (London: Dryden Press, 1886), 17; repr., 13. For a biography of Weigel, see Andrew Weeks, *Valentin Weigel, (1533–1588)* (Albany, NY: State University of New York Press, 2000). Weeks's biography agrees

"contained in none."[16] Furthermore, God is and abides most inward, which means that He inhabits all creation. That which is "exterior" is far from us, but God abides in us. Weigel's view of God is that of an occasionalist in that God is present at all times in the Creation and is in fact the underlying cause of visible and invisible phenomena. Should God cease to infuse nature, which would be counter to his being and desire, nature itself would cease.

Nature, the macrocosm, is ensouled by God. This soul is not God but has the Spirit of God in it. The material world is a body of fire, air, water, and earth. Like a body bereft of a soul, it is not dynamic and living. That soul is the Spirit of God manifested as the light of nature. Weigelians believed that "[t]he whole Light of Nature is in the soul of the Macrocosm."[17] They developed notions of the world's body and soul and their relationship to God without the obfuscation often encountered in mystical texts.

This soul of the world "hath in it the Spirit of God, which comprehendeth and possesseth it," for nothing is beyond God or the Spirit of God. Hence, the Spirit is nobler than the soul: "*The more noble always exists in the more ignoble, and internals prevail over externals, as in essence as in power.*"[18]

The world—matter and soul—is duplicated, as it were, in man the microcosm. Weigel believed that man is made of the parts of the universe. The Weigelians saw the universe as tripartite: the world of eternity, the "evial" world, and the world of time.[19] Likewise, "the parts of man are three," spirit, soul, and body. These parts of man were taken from the universe and, as in the universe, the spirit of God is integral and indwelling in man:

> The Spirit of man comes from the Spirit of God and participates with Eternity and *Ævo*.

---

with the role of soul knowledge in Weigel's Paracelsian-influenced theology. The *Astrology* can be treated as Weigel's for all practical purposes in the seventeenth century because that was then its attribution. John Warwick Montgomery, "Cross, Constellation and Crucible," *Ambix* 11 (1963): 75, treats the *Astrology* as Weigel's, as did Alexandre Koyré in *Mystique, spirituels, alchemistes du xvi siècle allemand* (Paris: Armand Colin, 1955), 81–116.

16 Weigel, *Astrology Theologised*, 14; repr., 67.

17 Ibid., 11; repr., 64.

18 Ibid., 12; repr., 66. The italics are in the English text and emphasize the point that the Spirit of God in man is a source of power.

19 "Evial" refers to something begun in time, but which is without end, such as the soul. Ibid., 10; repr., 63.

> The Soul in man is extracted from the soul of the World, and participates with *Ævo* and Time.
>
> The Body of Man is formed and composed from the body of the World, as elements, and participates with Time only.[20]

Because man is a little world or universe, man can comprehend within himself how the great universe functions. Nature is a manifestation of the divine. The Spirit of God is part of man and thereby is the vehicle of true knowledge. Weigel expressed the idea of soul-activated knowledge in terms of divine and natural wisdom sleeping in man. When awakened, man could understand physical, celestial, and divine processes:

> For both Divine and Natural Wisdom sleep in us, and each light shines in darkness, and without excitation, man wants the having.
>
> Great and excellent is the knowledge of the human body, extracted from the elements, and disposed into this form.
>
> Greater and more excellent is the knowledge of the Soul, taken from the firmament, and inserted into the body.
>
> Greatest and most excellent is the knowledge of the Spirit inspired from the Mouth of God into the first man, and by the mysteries of multiplication equally communicated to every one of us.[21]

Weigel rooted his belief that man's soul is the instrument of knowledge in the idea that looking inward reveals God's kingdom. He quotes Luke 17:21, "The kingdom of God shall not come with observation . . . for behold the kingdom of God is within you."[22] God is in us and therefore so is wisdom and truth. Weigelians expressed the doctrine thus:

> Wherefore, seeing there is nothing in us so near and intimate as God is, it follows that any other thing is not to be so esteemed, sought and loved as God alone, Who hath put and hid in us, the most excellent Treasure of His divine Wisdom, Light, Life, Truth, and Virtue, taken from His own Self, and

---

20 Ibid., 10; repr., 62–63.

21 Ibid., 10–11; repr., 63–64.

22 Ibid., 14; repr., 68.

> hath commanded to ask Him, seek, and knock in the hidden place of our heart, in Spirit, and in Truth, having given a testimony, that the kingdom of God, first of all, to be sought, is not here or there without us, but is to be found most inward in us as a Treasure hid in a Field.[23]

In the soul is collected "all kind of sciences, magic, Cabalistic, astronomic, with all their species, chemistry, medicine, Physic. . . ."[24] Soul knowledge could be accessed by theologizing the particular subject under study. This could be done, according to the Weigelians, by "more and more [seeing] the habitation of God in me."[25] To achieve this goal, one had to make study "sabbathized." To ensure divine favor and inspiration, every seventh day must be dedicated to God:

> Whatsoever therefore we have from the Light of Nature, all this with most humble self-denial once in the week is to be laid down at the feet of the best and greatest God, whether it be magic, or cabalistic, or astronomic, or chemic, or medicinal, or physical science.[26]

Divine wisdom, sleeping in the soul, could be unlocked by study and Sabbath worship. Weigel and his followers saw the study of the book of nature as being akin to the study of the book of scripture. First, one learned the text and then the Spirit of God within the human soul enlightened the mind to the correct meaning. Man could know, epistemologically speaking, because he was infused with the spirit and matter that he undertook to study.

Weigel's epistemology was logically derived from the macrocosm-microcosm analogy, Christian scripture, and his understanding of God's omnipresence. He claimed no particular originality. Like others inspired by Paracelsus, Weigel believed that the elements of his worldview were obvious to a Christian philosopher. His inspiration was Paracelsus who, in the Weigelian view, properly and completely described the creation of man as microcosm.

> Adam, the first parent of the whole human kind, was produced and formed by the admirable wisdom, and

---

23 Ibid., 15; repr., 70.

24 Ibid., 11; repr., 64.

25 Ibid., 17; repr., 72.

26 Ibid.

> workmanship of God, as to his Soul and body of the slime or dust of the earth; which slime or dust was such a mass or matter, which had conjoined and composed in itself the universal essence, nature, virtue and propriety of the whole greater world, and of all things which were therein. . . . The absolute description, and essential explication of this slime, dust or mass, extracted from the whole macrocosm, we shall find everywhere abundantly and wonderfully declared, alone by Theophrastus Paracelsus in his most excellent writings.[27]

Weigelianism, although theological, was consciously Paracelsian and hence part of the chemical philosophical tradition. The dominance of its theological orientation differed from that of the chemists because their primary focus was the study of nature. The Weigelian theological preoccupation allowed it to articulate the epistemological assumptions common to the Paracelsian chemists, but without systematic development by them.

III

Weigel's much younger contemporary, Joan Baptista Van Helmont, described the epistemology of illumination as it applied to his practice of chemistry. Van Helmont can be described as a Christian mystic who was especially influenced by the spiritual tradition of Pseudo-Dionysius, Hildegard, Thomas à Kempis, Renaissance Neoplatonism, and Christian Hebraism.[28] Unlike Weigel, he was not a trained theologian. He was a physician who wanted to develop a healing chemistry. His understanding of how one learns was focused on nature but was at the same time profoundly theological. He found his university education intellectually sterile and unfavorable to Christian learning: "I was diligent in the Art of *Logic*, and the Science Mathematical . . . the Study of Astronomy was of little or no account with me, because it promised little of

---

27 Ibid, 19; repr., 75–76.

28 Van Helmont, "The Author's Studies," in *Oriatrike*, 12. I have written about my view of the convergence of Christian Hebraism and Renaissance Neoplatonism in several articles. "Dee and Kepler," cited above, is representative of my position and notes other studies on the subject. Walter Pagel, in chapter 2 of *Van Helmont*, discusses Van Helmont's belief that the soul contains all the essences of the created world. This allows it to understand nature. Oddly, Pagel does not discuss the spiritual-illuminative aspects of Van Helmont's soul knowledge, which I explore here.

certainty or truth."[29] After reading Thomas à Kempis, he prayed to learn the "naked truth." Finally, he had a dream that convinced him "that in Christ Jesus, we live, move, and have our being."[30]

Van Helmont's epistemology grew out of his understanding of the centrality of God-Jesus to the Creation. In his "The hunting or searching out of Sciences," he notes that knowledge "begins from [Know thy Self]."[31] The ability to reason, to think syllogistically, might be accounted by some the very life of the soul, but "the Almighty is alone the way, the truth, the life, the light. . . ."[32] Van Helmont believed that the mind should be intellectual but not rational. All knowledge, whether natural or moral, proceeds from the self. The self is at root intellectual, but needs training in reason. Reason is a skill that can manipulate data but is not itself a source of truth, for it can be used to prove whatever one wishes. Van Helmont likens reason to a "thick and dark little Cloud."[33] It keeps the soul from truth. To properly learn and know, the soul has to free itself from reason:

> Also, now and then, Reason hath made Souls mad; who trusting too much to its perswasions, had enslaved themselves unto it.
>
> . . . The minde therefore hath drawn a wearisomeness from the command of Reason; and the rather, for that it knew Reason to be a Houshold Servant of its Family; but being a Chamber-maid, it presently did presume upon the whole government of the Soul.[34]

Van Helmont argues that mind and soul can escape from reason's hegemony through faith in God:

> And the minde having remembered that divine word, that those of his own House, are his enemies: conceived a loathing over Reason. And its turning away on both sides, was not yet sufficiently founded, yet it got strength in going From the first, therefore, after that, the Soul began no more

29 Van Helmont, *Oriatrike*, 12.

30 Ibid.

31 Ibid., 15. Brackets in original.

32 Ibid.

33 Ibid., 16.

34 Ibid., 17.

> to contemplate of Reason, as a part or power of it self.... And afterwards, the Soul knew that thing by faith, that it being once separated from the Body, it stands no longer in need of Reason.[35]

The soul is the proper instrument of knowledge because it is eternal: "*Solomon* [calls] the spirit of man, the lamp or candle of God."[36] The "Father of Lights" infused knowledge into the soul. The soul does not need nor does it use "Syllogisms" because it apprehends truth by its nature, just as it will when separated from the body in death. Van Helmont's introspection into the nature of the soul to gain certainty is similar to Descartes' viewpoint in his *Meditations.* Both are seeking a secure epistemological basis for knowledge. Van Helmont believed that the soul knew "things as they are," that is, their essence.[37]

Van Helmont believed that soul knowledge stands in stark contrast to the universities' teaching of Aristotle and the "heathens." He wrote of the correct epistemology "for the sake of Christians, for whom it is a shame, to follow Heathens contrary to Gospel truth."[38] The Gospel teaches—that is, Genesis teaches—that man was created in God's image. Our soul, which is the essence of that image, knows intuitively. The schoolmen ignore this Christian truth. They do not know man's nature:

> But the Soul of man, cannot be considered without the Image of God, seeing the Kingdom of God, with all its free gifts, is more intimate to the Soul, than the Soul it self is intimate to it self. Therefore, I am deservedly angry, that the Schooles do badly season Youth, with Heathenish Phylosophy, and truth, being now found. *Oh Lord, the light of thy cauntenance is imprinted upon us*; for none hath perfectly known the Image, and whether it doth well answer to its Type, if he shall not first know or acknowledge the

---

35 Ibid.

36 Ibid., 18.

37 Ibid., 20.

38 Ibid. Pagel, in chapter 3 of *Van Helmont,* discusses in detail Van Helmont's rejection of Aristotle.

> Type. Wherefore, as many as do badly define a man, do not know or acknowledge God. . . .[39]

Van Helmont could not easily describe the process of illumination that arises from the soul's experience of nature. It is intuitive, whether wholly from God or from the soul itself. Sense, reason, imagination, and memory melted into an insight or understanding. Van Helmont would rather partake of this "grace" than understand its operation.

> Then, then I say, a certain light falls upon the Soul. And that in my judgement, is all of whatsoever could ever be declared by word, thought, speech, and writing. But whether that light be altogether supernatural, or that the understanding be of its own nature thus kindled, or enflamed, I had rather experience than determine. That one thing at leastwise, I know, that it doth not happen without grace.[40]

Van Helmont describes a eureka moment for the Christian philosopher. This is the kind of insight that led Descartes to create analytical geometry; it is a sort of dream intuition. Van Helmont, however, believed that illumination could go beyond that found within the soul. The soul, through mystical union with God, could attain natural knowledge directly and immediately from God. This could be accomplished through a Kabbalistic process described by the rabbis. Van Helmont seems to have derived his belief in this intense union with the Divine from Pico's articulation of the "mori ex binsica," or kiss of death.[41]

---

39 Van Helmont, *Oriatrike*, 21. Allen G. Debus, *Chemistry and Medical Debate: Van Helmont to Boerhaave* (Canton, MA: Watson Publishing, 2001), 32–51, discusses Van Helmont's search for truth, his rejection of an astronomical model for medicine, his religious quest, and dreams. Of special interest for this study is Debus's interpretation of the tomb of true medicine, which serves as frontispiece of Van Helmont's *Aufgang der Artzney-Kunst*, as a representation of Van Helmont's dream. Debus also notes that, on the basis of Genesis, Van Helmont determined that fire is not an element.

40 Van Helmont, *Oriatrike*, 23.

41 Pagel, *Van Helmont*, notes Van Helmont's belief in the divine kiss (24). Pico's use of the idea and its sources are discussed in Wirszubski's *Pico della Mirandola's* (58–59). Nicolas J. Perella discusses the *mors osculi* in *The Kiss Sacred and Profane* (Berkeley: University of California Press, 1969). He attributes its origin as a mystical illumination to Pico (180). The kiss as a means of death for the righteous was well known from the passage in Deuteronomy 34:5. Among others, Leone Ebreo, J. Reuchlin, H. C. Agrippa, and Giordano Bruno saw the

With the "kiss of death" the philosopher offered his soul to God, which when accepted brought about an intellectual rapture and divine illumination. The concept's origin was in rabbinic commentary on Deuteronomy 34:5, the scriptural description of the death of Moses. The passage states, "And Moses, servant of the Lord, died there, in the land of Moab, by the mouth of the Lord." Rashi, the eleventh-century Jewish biblical commentator par excellence, following Talmud Bavli *Moed katan* 28a, wrote, "'by the mouth of the Lord' [means] by means of a kiss—*binishicah*." In Hebrew, *Nishicah* is a kiss. *Bi* is a preposition meaning "by means of." Jewish Kabbalists came to associate the divine kiss with mystical union and illumination, rather than just as a means of death unique to Moses. The thirteenth-century Kabbalist Menachem Recanati made no distinction between actual death and the soul's mystical adherence to God. This attitude toward the *Nishicah* is precisely that of Pico in his Kabbalistic theses.[42] Flavius Mithridates, who translated Hebrew manuscripts into Latin for Pico, was highly influenced by Recanati. His translation of Recanati's *Commentary on the Pentateuch* was a primary source of Pico's Kabbalah.[43]

Pico passed on the Kabbalistic interpretation of the *binsica* [his spelling, also used by Van Helmont] to the scholarly world in his Kabbalistic thesis 13.[44] This death, according to Pico, was not the external death or the body's separation from the soul, but rather a rapture in which the soul separated from the body ("non est nisi per separationem animae a corpore, non corporis ab anima").[45]

Van Helmont's belief that light from God comes from an interior "reflexum" pushed him to go beyond the inner light to its source as described by Pico. ("Eo quod hoc cognitiones lumen, non sit emanativum extra intellectum: sed manet intus reflexum. . . .") This knowledge within the soul is

---

kiss as an "ecstatic death" and a "mystical rapture" during contemplation (181). Edgar Wind, *Pagan Mysteries in the Renaissance* (New Haven, CT: Yale University Press, 1958), traces the Renaissance association of love and death to the ancient Neoplatonists' interpretation of the embrace of Psyche and Eros (33 and 134). Van Helmont seems to be original in adapting the concept to natural knowledge.

42 Wirszubski, *Pico della Mirandola*, 58–59.

43 Ibid., 10–21.

44 Ibid., 159.

45 Ibid., 158.

superior to the intellect and is "in all ways true and perfectly certain [super intellectum, in omni veritate, et certitudine perfecta]."[46]

To arrive at this perfect certainty was dangerous and required following the pathway of the mystic and the adept. It was accomplished through the "undoubted death" with the "affliction" that the rabbis call the "Binsicam." This illumination is the highest form of knowing and should be the goal of the true reader of the book of nature. Walter Pagel first noted this crucial aspect of Van Helmont's epistemology. It seems reasonably clear from the text that Van Helmont was not simply stating an ideal, but describing what he regarded as a necessary step in the perfect study of nature. Van Helmont believed that he had reached this level and that it was the source of his greatest insights.

IV

Johann Joachim Becher (1635–82) was the son of a Lutheran minister. His chemical education seems to have been picked up during his travels through Europe. His career included being a court physician and an imperial commercial counselor. He is best remembered as a chemist and as the primary source for Georg Ernst Stahl's phlogiston theory.[47] His *Psychosophia* or "Soul Wisdom" explicates and summarizes his conception of illuminative epistemology. The book was meant to be both a natural philosophical treatise and a contribution to the economic development of Europe. Becher believed, as had his chemical predecessors, that the soul is the ultimate source of knowledge. His understanding of chemistry and its mission to help mankind was rooted in the chemist's attainment of illumination or soul wisdom.

---

46 Van Helmont, *Ortus Medicinae*, 3rd ed., 24. For purposes of greater precision, I translate from the Latin rather than using the seventeenth-century English translation, *Oriatrike*, that I have uniformly quoted above.

47 The most accessible discussion of Becher's career is P. H. Smith, *The Business of Alchemy: Science and Culture in the Holy Roman Empire* (Princeton, NJ: Princeton University Press, 1994). Phlogiston was a substance which entered or left bodies during heating. Metals combined with phlogiston when heated and became lighter. When reheated, phlogiston would leave the metal making it heavier. Lavoisier did away with phlogiston by identifying oxygen as the agent of change and proving that metals gain weight in heating or calcination.

Psychosophia oder Seelen-weissheit describes how "every person could attain all science (wissenschaft) and wisdom (Weissheit) solely through contemplating his soul." The frontispiece (*fig. XI*) illustrates the process of soul wisdom and how it works.[48] A man sits looking into a mirror. Just outside the mirror is a small, yet wingless, cherubic personage who represents the soul. Above is a three-letter representation of the Tetragrammaton, with rays going to the soul and to the mirror. Becher explains the frontispiece in a poem placed before the preface:

> The mirror represents the Soul
> When God illuminates the same
> So sees Man therein
> Himself portrayed [his soul]
> When, however reason attempts to illuminate the Soul
> It is uncertain
> And it [the soul] often remains very still [is imperceptible].[49]

In the first part of the *Psychosophia*, Becher presents a dialogue between a philosopher and a "psychosopho." The philosopher asks, "Who are you?" The psychosopho answers, "I am a man." "Of what are you composed?" "Of body and soul." The questions and answers establish that man is made of the four elements that form the vital organs and blood. The soul is a divine creation. As in the works of Weigel and Van Helmont, this relationship of soul to body is the basis of the idea that the material world can be known through the soul. The dialogue form highlights the epistemological limits of philosophy and the epistemological strengths of psychosophia. Becher testifies that what he wrote [and published at his own expense] "is true" and not from "other books" but rather from his "own head."[50]

Becher declares that he did not write the *Psychosophia* as a theoretical book but rather to present the epistemology to be used by a Christian scientific society, which he hoped to establish. That society would not be divided by sectarian theologies but would work to benefit mankind through

---

48 The edition used for this essay is the second (Frankfurt, 1683). All translations mine.

49 Ausslegun des kupffer Rituls:
Der Spiegel deut die Seel/ wann Gott dieselb bestrahlet/ So sieht der Mensch darinn/ sich selbsten abgemahlet/ wann aber die Vernunfft die Seel bestrahlen will/ So ist es misslich/ und steht oeffters gaentzlich still.

50 Becher, *Psychosophia*, preface, fol. ii verso.

Figure XI. Title page. Johann Joachim Becher, *Psychosophia oder Seelen-weissheit*, 2nd ed. Frankfurt, 1683.

chemical studies broadly understood. The society would be rural and agrarian. Technologies would be developed through psychosophia to improve agriculture and urban life.[51]

The society's research would be done following the chemical understanding of matter and using chemical processes. Chemistry taught that the philosophers' stone could heal illness. Metals can be purified and mixed substances, such as glass, can be improved.[52] Psychosophia underlay the entire program of the society. Healing tinctures, philosophical mercury, and the philosophers' stone could not be made or analyzed without the true illumination of soul study. Alchemy [chemistry] was based on nature and nature is understood through experience and the light of truth within the soul. The philosopher explains to the psychosopho:

> I see now why one calls the tinctures universal, because they offer a universal understanding and scientific knowledge of natural things. . . . I have seen a great light of the truth arise from your speeches, [namely, the limits of philosophy and the deep insights of psychosophic chemistry].[53]

Becher saw psychosophia or soul wisdom as the key to nature. That key would be of most value if used in a social-scientific context. The best venue, therefore, was a Christian society of chemists united in the study of nature working together to benefit humanity. One could best gain soul knowledge or illumination through experiencing nature as a Christian chemical psychosoph in the company of others in that holy order.

For Weigel, Van Helmont, and Becher, the recognition of the soul, God's image in man, was the epistemological preface to experiencing nature through chemical studies. It is by means of experience in the laboratory that the soul encounters the data necessary for intuition to operate. How this epistemology worked in the laboratory is illustrated in the many chemical experiments or ocular demonstrations mentioned by Robert Fludd and Van Helmont.

---

51 Ibid., preface, fols. iiii–v.

52 Ibid., 158–59.

53 Ibid., 165.

V

Sixteenth- and seventeenth-century chemical experimentation, as noted in the previous chapter, differed from the modern idea of making experiments. Fludd and Van Helmont conceived quantitative demonstrations that would reveal natural processes. These demonstrations were an important part of the evaluative procedure, as was reproducibility. Whereas modern experiments are performed to test physical theories, Fludd and Van Helmont sought to test, or to state it more clearly, to illustrate, both spiritual and physical theories. This twofold function of experimentation was in large measure the result of the illuminative epistemology. The chemical experimental method is illustrated in Fludd's study of wheat and Van Helmont's analysis of ferments.

Robert Fludd's *A Philosophical Key* (ca. 1619–20) was a study of wheat written as an "[o]cular demonstration, opening and decyfering a great deale of the hidden mysteryes of Nature, partly by an experimental Conclusion, as also by an intellectual speculation."[54] Fludd studied wheat because it was the king of vegetables, for the nourishing mother nature was most resident in it. Fludd believed that the "pretious balme of Nature wch is derived from the bright essence of the heavenly sonne" is hidden in wheat and is responsible for the plant's life and "multiplication." Fludd describes wheat's growth from "putrefaction" in the earth until its "flight" upward—a process like resurrection. The divine spark comes from the Sun and enters dead matter, causing it to live. The life spark then rises toward its solar source:

> . . . then will it as I guess be very easy for us to know why the eare of corne boulteth upright out of the earth and doth (as it were) rayse himselfe on the top of a straight slender and pinie trunke: for after putrefaction is made (without which the scripture: teacheth: us that nothing wch is deade can rise agayne or be exalted) that never dying sparke of the fifte [fifth] essence, being im̃ediatly derived from the heavenly lyne of that sacred Appollo who hath built his tabernacle in

---

54 Robert Fludd, *Philosophical Key*, in Allen G. Debus, *Robert Fludd*, 63. The *Philosophical Key* was in part written (but not published) as a response to criticism of Fludd's *De Macrocosmi Historia* and his defense of the Rosicrucians. It is a defense of his philosophy and his experimental (experiential) methodology. Debus's introduction provides an extensive context for Fludd's work.

> the sonne, finding it selfe now as it weare free through putrifaction, from the bonds of the fower elements which kept him in his compound prison: beginneth to take his flight towards the Æthereall region from whence it came; and by litle and litle exalting his creste above the darke and heavye earth, the element of fire (which by a wonderous inclination and inviolable bande is anexed unto that quintessentiall spirit) by subtill fumes and exhalations climeth up immediately.[55]

Fludd proceeds to analyze wheat, finding that it is mostly water, which it had sucked from the earth. It also contains a subtle "aery" element with many "hidden facultys," "fyre," and an "aethereall" spirit. The subtle aethereal or quintessential spirit is that which sustains life in man. Of course, wheat nourishes man when it is baked into bread. Bread is mentioned as life-giving in the scriptures. Adam was told, by the sweat of your brow you will eat bread.[56] For Fludd, there are two kinds of bread, both of which come from God:

> Of two Kinds of bread wherof the one is terrestriall and vulgar, the other coelestial, & proceedinge im̃ediatly from God. And of the marvelous virtue . . . wch it hath wth yt wch is spirituall.[57]

Fludd discusses the importance of the two breads and then reports a series of experiments or "apparetions" in which he subjected wheat to chemical analysis. He wanted at first to separate the volatile elements from the "fixer mass." He sought to "ape" nature and learn the nutritive secrets of the plant. At the same time, Fludd felt that he was glimpsing the processes of creation found in Genesis. His analysis of wheat, therefore, became a commentary on Genesis. His soul illuminated him both as to practical nutrition and as to theological truth. He clearly saw that water is the basic substance of all things:

> The subterrestriall agent of the Author: The comparisons of thos apparitions in his first vision, with thẽ of the universall creation. An evident proofe yt all things are water. The

---

55 Ibid., 94.

56 Fludd quotes the Hebrew.

57 Ibid., 101, chapter heading.

> reason why the worlds matter is grosser in one place then in an other. The description of his first vision which he saw and the relation betwene it and the principall effect in the greater worlds creation:[58]

The interplay of chemical analysis, scripture, and illumination typical of chemical philosophers is readily apparent in Fludd's understanding of experiment.

Van Helmont represents a more rigorous physically analytical strain of chemical philosophical experimentation. His discussion of ferments is at once critical and spiritual in that it arises from his belief that water, the basic element of Genesis, is the source of all physical things. It is also a major point on which he attacks scholastic philosophy.

Van Helmont begins his discussion of ferments by stating, "[N]o knowledge in the *schooles* is scantier than the knowledge of a Ferment."[59] The learned world limited the concept of fermentation to the leaven used in making bread and brewing beer. In fact, Van Helmont believed, ferments and ferments alone bring "successive change or transmutation of matter."[60] As did Fludd, Van Helmont associated corruption or putrefaction with the process by which a ferment or leaven transmutes matter. He demonstrates the action of fermentation by analyzing the brewing process:

> That *Ale* or *Beer*, by a very little *ferment* or *leaven* being administered, doth boyl up by fermenting in Cellars, it waxeth clear by degrees, and the dreg falls down to the bottom. . . . At length it is deprived of the taste, virtues, and body of the meal. And last of all, it, of its own accord, returns into *water*. . . . Therefore the Grains do return unto their first matter whereof they are, which is water, and that by the virtue of the ferment onely.[61]

Van Helmont understood that beer could not be made without a large quantity of grain. Yet, a minute amount of ferment brings significant changes that even distillation cannot accomplish. Ferment reduces what would have

---

58 Ibid., 112, chapter heading.

59 Van Helmont, *Oriatrike*, 111. Pagel, in *Van Helmont*, discusses ferments and their images, odors, in detail (71–74).

60 Van Helmont, *Oriatrike*, 111.

61 Ibid., 112.

been dregs in distillation to their first matter, water. Van Helmont further illustrates the indispensability of ferments by pointing to the blood. He believed that the blood is a nutrient that is produced and consumed daily. Ferments are responsible for turning blood into the substance that nourishes the body in much the same way that the nutritious essence of grain is produced by fermentation:

> In the next place, every one of us doth daily frame to himself, 7 or 10 ounces of bloud; but (at leastwise in our standing age) as much bloud must needes be consumed, as is a-new, generated: For else a man might straightway fear a hugeness or excessive greatness. And then, the bloud is by degrees, changed into a vitall muscilage or slimy juyce, the true, immediate nourishment of the members of which it is wont to be said, *we are nourished by those things, whereof we consist*. But they will have this nourishment to be sprinkled on all the particular members; in manner of a dew (but I believe it to be framed in all the least Kitchins of the parts) whereby it may moysten the same, and for that cause, defend them from dryness, the calamity of old age, as much and as long as it can. At length, that dew doth unperceiveably flee thorow the pores of the skin, neither doth it leave any thing of a solid sediment remaining behinde it: For so do nourishments at length exspire thorow the skin in the shew of a Vapour, and like water.[62]

The blood is changed by ferment in the kitchens of the parts. This metabolism, which produces vital moisture, sustains life and youth. The vital moisture eventually dissipates and old age ensues. The "vapour" that passes off through the pores is like water. In the analysis of all things, Van Helmont and Fludd returned to water as the substrate from which every thing arises and into which they return. Van Helmont considered water fundamental for two reasons. First, the use of water is fundamental in experiments, such as those involving brewing, the blood, and especially the tree's growth, which is noted above. Second, water is the prime matter of Genesis. The Spirit of God, like a ferment or ferments, created the particulars of the world from the primordial water. Remember, Van Helmont had written that "all plants arise imme-

---

62 Ibid., 112.

diately and substantially from the element of water alone."[63] Ferments—activating the principles of seeds—are the "parennts of [such] transmutations." The chemist "by the art of fire" could explore the transmutation of bodies, understand through soul wisdom the creative processes, and learn to use them. Van Helmont combined Genesis with chemistry in order to understand the chemical processes of ferments, something essential to using nature, which was little known in the schools, but obvious to students of the chemical philosophy.

[63] I discuss this at length in "Genesis and Chemistry in the Sixteenth Century," in Debus and Walton's *Reading the Book of Nature*, (12).

6. Genesis and Chemistry to the End of the Seventeenth Century

This book began with a discussion of the theological interpretation of Genesis from the ancient world to the Renaissance. Genesis is the divinely revealed description of the Creation. Early Christian theologians believed that the creation story in Genesis had to be harmonized with pagan natural philosophy. The doctrine of the elements and pagan philosophical points of view became part of Jewish and Christian readings of Genesis. During the Renaissance, there was a convergence of ideas from the ancient world, Jewish, Gnostic, Neoplatonic–Neo-Pythagorean, which seemed to prove that all correct knowledge originated in God's revelation to man from Abraham to Moses, the *prisca theologia* or *doctrina*.

Renaissance Christians wanted to recover the true philosophy from the encrustations of pagan falsehood and erroneous Christian understandings. Luther sought the original meaning of scripture and became a reformer par excellence; however, he still used pagan elemental theory in his explanation of the creation. Alchemy was another interpretive tradition of the creation. It did not place matter theory and chemical processes into the service of theology, but, rather, placed theology into the service of the "art." The search for the "stone" of transmutation was both a physical and a spiritual one. The alchemical tradition's origins in Gnosticism and practical chemistry made it easy to connect with the Genesis account of Creation, on the level of analogy. The Genesis account and its interpretation converge with elements of alchemy in several crucial respects. Distillation was seen as an analog of Genesis that the adept could observe. Water and its changing states were part of that analogy, as was the idea that the macrocosm and its processes existed in the microcosm, man.

The alchemical tradition was "transmuted" by Paracelsus in his search for a Christian reading of the book of nature. Paracelsus marked the beginning—at least in the eyes of his successors and opponents—of the chemical philosophy. Not all practitioners of chemistry followed that "Christian" way of understanding and doing chemistry. The non- or anti-Paracelsian chemists are not the subjects of this study. The Paracelsian-chemical philosophy was enormously influential and controversial for more than a century. Paracelsians—Dorn, Croll, and Severinus—systematized the master's thought and established it as theologically acceptable because it had been developed from Holy Writ. They believed that work in the laboratory would produce remedies that would revolutionize medicine.

As the philosophy developed, Khunrath and Fludd placed it within the context of Kabbalah, Neoplatonism, and chemical experimentation. Van Helmont, Glauber, and Becher were experienced chemists who went beyond Paracelsus and criticized him. Still, they acknowledged that he was a crucial figure in the development of a chemistry freed from pagan ideas and the scholastics. Believing in the importance of the literal description of Creation given in Genesis was the point of departure for Paracelsus and the chemical philosophers who followed him. In their own eyes, they were no longer dependent upon pagan traditions. They saw that Genesis could give them the key to unlock the book of nature. The creation account contained valuable chemical knowledge. Their matter theory and understanding of chemical processes began with Genesis.

These diverse personages and their idiosyncratic points of view, which over time shaped the chemical philosophy, saw nature as the veil that hid God from man. Yet, God had given them the ability to draw back the veil. The proper reading of nature would reveal the Creator and, as a gift from him, give insight into medical remedies and other wonders, to the benefit of mankind.

The theologian-chemical philosophers, such as Bostocke, Weigel, and Tymme, were attracted to an epistemology dependent on spiritual illumination. Had Adam not fallen, man's unstained soul would be able to see the Creator and his creation immediately and unmediated. Yet, the honest student of theology and chemistry could apprehend truth; the soul could find again that which had been revealed to Adam, Abraham, and Moses.

Paracelsus, Dorn, Croll, Khunrath, Fludd, Bostocke, Weigel, Becher, Van Helmont, and Glauber all testified in their writings to the revelatory character of natural philosophical knowledge.

The shared illuminative epistemology of chemical philosophers was based on the general Christian idea that the spirit of man shares the same origin as nature, both are creations of God. This epistemology is closely related to the Christian notion of personal enlightenment by the Holy Spirit as at Pentecost and with St. Paul on the road to Damascus. Illumination or soul knowledge was placed in opposition to the possible circularity and vanity of pagan logic. This vision of Christian knowledge led to a methodology of personal illumination through experiment. Chemists did not believe that the mathematical abstractions or *resolutio* and *compositio*, with their emphasis on causality, were the correct way to understand the dynamic processes of life and chemistry. Christian philosophers were better served by allowing their souls to experience nature and apprehend the truth of its processes as seen in the laboratory and the great outside world. Genesis is a guide to soul knowledge in that it shows God infusing himself into the Creation and creating man (male and female) in his image.

Genesis was essential to the chemical philosophy from the middle of the sixteenth century into the seventeenth century. As with all intellectual movements, the chemical philosophy declined in its usefulness and interpretive vigor. Needless to say, chemists did not become less religious, or less Christian; rather, what the interpretation of Genesis had to offer to chemistry became rooted in theory and practice.[1] It did not need to be restated. Experimentation focused on physical factors and not on theology. Chemistry had become "Christian science." It became a self-sufficient, physical reading of the book of nature.

---

1 Several works have chronicled and discussed the development of chemistry from the seventeenth century into the eighteenth century. This book makes no pretense of being a history of early modern chemistry. See Debus's *The Chemical Philosophy*, *The French Paracelsians*, and *Chemistry and Medical Debate*. See also Principe, *The Aspiring Adept*; Newman and Principe, *Alchemy Tried in the Fire*; and Bruce T. Moran, *Distilling Knowledge: Alchemy, Chemistry and the Scientific Revolution* (Cambridge, MA: Harvard University Press, 2005).

In the theological arena, the Genesis account of Creation and science continued their association through the seventeenth century. In England, the hexaemeron was represented throughout the seventeenth century by John Swan's popular *Speculum mundi ... occasioned as matter pertinent to the work done in the six dayes of the world's creation* (Cambridge, 1635). Swan, a doctor and cleric, evinced no particular originality of scientific knowledge; his focus was primarily theological. The *Speculum* was published five times, the last printing being in 1698.

The theological interpretation of nature reached its apex in the work of the so-called Cambridge Neoplatonists, Ralph Cudworth and Henry More. They used science as the early Christian exegetes had done, to prove that nature declares God's glory. In 1678, Cudworth published the first part of his *True Intellectual System of the World.* (The work in its entirety was not printed until 1705.). In this dense volume, Cudworth refutes atheism, that is, materialistic philosophy. He demonstrates that atheism is not a "necessary appendix to atomism." Pierre Gassendi and Walter Charleton had Christianized atomism and made it a part of physics and chemical theory in the mid-seventeenth century.[2] Demonstrating his immense learning, Cudworth analyzes philosophy from Moses to his own age and proves that the correct natural philosophy is based on a Christian spiritualized or Neoplatonic atomism.[3] Cudworth's work was theologically influential into the nineteenth century.

More important for chemistry, perhaps, were the physico-theological notions of Henry More (1614–87). More, like Cudworth, adhered to the belief that Plato knew the works of Moses and had based his philosophy on that knowledge. Had not Numenius called Plato "Moses Atticus"?[4] More's

---

2 The "baptism" of atomism by Pierre Gassendi and others is discussed in Sailor's "Moses and Atomism." Robert Kargon's introduction to the reprint of *Physiologia Epicuro-Gassendo-Charltoniana* (London, 1654; repr., New York: Johnson Reprints, 1966) and Antonio Clericuzio's *Elements, Principles and Corpuscles: A Study of Atomism and Chemistry in the Seventeenth Century* (Dordrecht: Kluwer, 2000) also give valuable information on the topic.

3 The text I have used is the first complete American edition of Birch. Ralph Cudworth, *The True Intellectual System of the Universe, with References, and the Author's Life by Thomas Birch* (New York: Andover, 1837).

4 Henry More, *Conjectura Cabbalistica* (London, 1653), 13.

Conjectura Cabbalistica was not in the Judeo-Christian Kabbalistic tradition of Pico, but rather an exposition of the literal, philosophical, and moral meaning of the creation story in Genesis. It sprang from More's own philosophy and not from Hebrew mystical texts.

More used traditional notions that were already well represented in the chemical philosophy. Wisdom created the world by means of spiritual substances and seminal forms. The spiritual made actual the dark inchoate possibility:

> And God approved of, and pleased himself in all this as good; but yet though in designe there was a settlement of the *fluid darknesse* or obscure *Possibility* of the outward Creation, yet it remained as yet but a dark Possibility.[5]

More's theological preoccupation naturally led him to use science to prove his beliefs. His lack of technical scientific knowledge brought him into conflict with his friend Robert Boyle. This conflict illustrates that serious chemical practitioners working in the laboratory, such as Boyle, were coming to see their discipline as distinct from theology, having its own self-contained method and rules of interpretation. But More, like Cudworth, thought it was necessary, in order to understand nature, to impose spiritual forces onto the mechanical ideas fostered by corpuscularianism or atomism. As a result, he suggested in 1662 that spiritual substance is at work in the behavior of gases:

> A substance incorporeal, but without Sense and Animadversion, pervading the whole Matter of the Universe, and exercising a Plastical power therein according to the sundry predispositions and occasions in the parts it works upon, raising such *Phaenomena* in the World, by directing the parts of the Matter and their Motion, as cannot be resolved into mere Mechanical powers.[6]

Boyle, who was deeply religious and a student of Hebrew, and who had described, in *The Sceptical Chymist*, how the Spirit of God incubated the waters in order that the "component parts" of the world could emerge, reacted

---

5 Ibid., 24.

6 Henry More, *The Immortality of the Soul* in *A Collection of Several Philosophical Writings of Dr. Henry More* (London, 1662), 193/[579].

negatively to More's meddling in scientific matters.[7] Although Boyle, like Van Helmont, believed in seminal forces, he did not see the need for spirit forces. He wrote, "[O]ur Hydostaticks do not need it."[8] Boyle began to set limits on chemistry, confining it to what was experienced or could be replicated in the laboratory. This is shown by his pulling back from using Genesis in *The Sceptical Chymist*. After describing water as the prime matter and the action of the Spirit of God, he quickly states that he should discourse as a naturalist, not as a theological interpreter of biblical language:

> [B]y perusing the beginning of Genesis, where the waters seem to be mentioned as the material cause, not only of sublunary compound bodies, but of all those that make up the universe; whose component parts did orderly, as it were, emerge out of that vast abysse, by the operation of the Spirit of God, who is said to have been moving Himself, as hatching females do (as the original, *Merahephet*, is said to import, and it seems to signifie in one of the two other places, wherein alone I have met with it in the Hebrew Bible) upon the face of the waters; which being, as may be supposed, divinely impregnated with the seeds of all things, were by that productive incubation qualified to produce them. But you, I presume, expect that I should discourse of this matter like a naturalist, not a philologer.[9]

Boyle's reluctance to derive science from theology did not mean that theology was unimportant to his chemistry. Quite to the contrary, he was a theological man through and through. Nonetheless, he argues that the scientific method and rules of proof, which are experimental and physical, differ from theology. He was both a theological literalist and a scrupulous chemist:

> I see no just reason to embrace their opinion, that would so turn the two first chapters of *Genesis*, into an allegory, as to

---

7 Robert Boyle, *The Sceptical Chymist*, Everyman's Library, Science, no. 559 (London, 1661; repr. London: Dent, 1911), 71–73. See Michael T. Walton, "Robert Boyle, 'The Skeptical Chymist' and Hebrew" in Williams and Gunnoe, *Paracelsian Moments*, 187–201.

8 Robert Boyle, *A Hyrdrostatical Discourse occasion'd by some Objections of Dr. Henry More* (London, 1672).

9 Boyle, *Sceptical Chymist*, 71–72; see also my "Genesis and Chemistry in the Sixteenth Century."

> overthrow the literal and historical sense of them. And though I take the scripture to be mainly designed to teach us nobler and better truths, than those of philosophy. . . .

Whatever his own opinion, he understood that others still looked to Genesis for enlightenment:

> [Y]et I am not forward to comdemn those, who think the beginning of *Genesis*, contains divers particulars, in reference to the origin of things, which though not unwarily, or alone to be urged in physicks, may yet afford very considerable hints to an attentive and inquisitive peruser.[10]

Thus Boyle represents a change in the use of Genesis by chemists. He repeated the experiments that demonstrate "seminal" action on water, but he did so without reference to the Genesis text. This approach is also found in Newton, in both his *System of the World* and his *Opticks*. He used Helmontian and Boylean ideas, but he provided little chemical and minimal natural theological context.[11] Newton was profoundly influenced by theology and by mystical chemical texts, but he largely avoided mention of these in his published scientific writings.[12] When the Royal Philosophical

---

10 Robert Boyle, *The Excellency of Theology or the Pre-eminence of the Study of Divinity above That of Natural Philosophy*, in *The Works of the Honourable Robert Boyle*, ed. Thomas Birch, 6 vols. (London: W. Johnson, et al., 1772), 4:11; see also Walton, "Robert Boyle, 'The Skeptical Chymist' and Hebrew," 202–3.

11 Walton, "Boyle and Newton." Newton's decision not to publish his "Classical Scholia" to the second edition of the *Principia* illustrates the narrowing of natural philosophy's universe of discourse. The "Scholia" refers to the *prisca*, clearly something in which Newton believed. In the last analysis, however, he seems to have seen such discussions as outside of science. See J. E. McGuire and P. M. Rattansi, "Newton and the Pipes of Pan," *Notes and Records of the Royal Society of London* 21 (December 1966):108–43. I. B. Cohen, in "Isaac Newton's *Principia*, the Scriptures and Divine Providence," *Philosophy, Science and Method*, ed. Sidney Morgenbesser, et al. (New York: St. Martin's Press, 1969), 523–48, notes that Newton's brief reference to God in the first edition of the *Principia* was removed by him from the second and third editions. The "General Scholium" of the second edition is Newton's primary printed theological statement. It is presented in such a way as to distinguish it from the science of the work. The 1704 first edition of the *Opticks* does not contain overt statements of natural theology, but Newton added four such discussions to later editions.

12 Betty Jo Teeter Dobbs, *The Foundations of Newton's Alchemy or "The Hunting of the Green Lyon"* (Cambridge: CambridgeUniversity Press, 1977) and *Alchemical Death and*

Society was established in 1660, it excluded studies of metaphysics and divinity, as well as other topics. This further demonstrates the limitations of the scientific universe of discourse seen in Boyle and Newton. By the end of the seventeenth century, science had been established as a discipline distinct from theology, so much so that the revelation of Newton's theological predispositions in the eighteenth century by his student William Whiston was profoundly disturbing to both scientists and theologians.[13]

In the sixteenth century, chemical philosophers challenged the universities, claiming they were bastions of paganized scholasticism, where Christian chemistry was unwelcome. The fact that most chemists were university trained would seem to give some credence to their arguments; they had firsthand knowledge. By the mid-seventeenth century, however, the universities had changed considerably with regard to their approach to natural philosophy. The accomplishments of Copernicus, Kepler, and Galileo affected the teaching of the mathematical sciences; chemical discoveries affected medicine. The new science and the chemical philosophy were no longer foreign to academe.

In England, this change was the subject of debate between the nonconformist cleric John Webster (1610–82) and mathematician divines John Wilkins (1614–72) and Seth Ward (1617–89). Webster attacked the academies from a chemical philosophical position in *Academiarum Examen* (London, 1653). He argued that although universities were necessary, they were mired in old-fashioned ideas based on pagan-popish philosophy. They were failing their students and their duties by neglecting to address the spirit of the Gospel and to teach godly things. They taught in popish Latin, when

---

Resurrection (Washington, DC: Smithsonian Institution Libraries, 1990); and Matt Goldish, *Judaism in the Theology of Sir Isaac Newton* (Dordrecht: Kluwer, 1998). Goldish reproduces several important theological manuscripts of Newton. Stephen D. Snobelen, "Isaac Newton, Heretic: The Strategies of a Nicodemite," *British Journal of the History of Science* 32 (1999): 381–419, discusses Newton's concealment of his theology.

13 James E. Force, *William Whiston: Honest Newtonian* (Cambridge: Cambridge University Press, 1985), details the "Newtonian religion" and its conflict with what became Newtonian science.

they should be teaching in the vernacular, and were ignorant of Christian modern science.[14]

Webster's charges did not go unanswered. John Wilkins, warden of Wadham College, had earlier defended Copernicanism and Seth Ward was a budding astronomer and mathematician who in 1656 would defend the ideas of Kepler. In their *Vindiciae academiarum* (Oxford, 1654), they pointed out to Webster that the universities already taught, in large measure, what he advocated. Moreover, they said, Webster's point of view was dated and his ideas, in many ways, imprecise and incomprehensible. They recommended that Webster learn about the universities and their curricula before attacking them.[15]

The Webster-Ward debate highlights the rhetorical and factual gulf between the claims of the chemical philosophy and the intellectual realities at the universities by the middle of the seventeenth century. While throughout England and Europe, there would be empirical and nonuniversity trained chemists and even alchemists, the universities increasingly became the places where the new sciences of astronomy, mathematics, and chemistry were developed. The need to alter or overthrow the scholastic system with a Christian philosophy was declining. Official pharmacopoeias contained chemical remedies; the best medical schools had some kind of chemical component; and students were taught the mathematical astronomy and physics of Copernicus, Kepler, and Galileo.

At the beginning of his astronomical career, Kepler himself had been inspired by a divine vision, which motivated him to pursue the study of planetary motion. In that pursuit he did not rely on revelation. He worked out his *Astronomia Nova* (1609) with demonstrable mathematical proofs.[16] He, Boyle, and Newton represent theological men who practiced a kind of science that assumed theology but used rigorous physical-mathematical demonstrations. Genesis, as it were, was still spiritually present, but was less likely to be explicitly referred to on the printed page.[17] The recognition that

---

14 Allen G. Debus, *Science and Education in the Seventeenth Century: The Webster-Ward Debate* (London: MacDonald, 1970), 37–43.

15 Ibid., 43–51.

16 Michael T. Walton and Phyllis J. Walton, "The Geometrical Kabbalahs."

17 Williams argues that technical Bible scholars, such as DeLaHaye, *Biblia Magna* (1644), and John Pearson, *Critici Sacri* (1660), turned from using science to explicate the text: "So

science is in the province of the naturalist and not in that of the "philologer" marked the rapid decline of Genesis as an explicit datum of physical theory; soul knowledge and illumination were no longer an explicit epistemology for mainstream natural philosophers. Still, students of nature did not become demonstrably less religious or agnostic. In the sixteenth and seventeenth centuries, the growth of the Christian chemistry served to help develop chemical and epistemological theories that resulted in ideas and methods that bore fruit for more than a century and became essential aspects of the scientific method.[18] While scientists refrained from invoking scripture, theologians used science after the manner of St. Basil to prove religion.

---

far as I have discovered, succeeding exegetes were satisfied with the divorce between hermeneutics and science" (Williams, *Common Expositor*, 262).

18 The chemical, Neoplatonic, Neo-Pythagorean, Kabbalistic worldview influenced German pietism through figures such as Jacob Boehme, F. M. Van Helmont (the son of Jan), Christian Knorr von Rosenroth (*Kabbalah Denudata* [Sulzbach, 1677]), and F. C. Oetinger. See Ernst Benz, *Christian Kabbalah*, trans. K. W. Wesche (St. Paul, MN: Grailstone Press, 2004); Allison P. Coudert, *The Impact of the Kabbalah in the Seventeenth Century* (Leiden: Brill, 1999), and *Leibnitz and the Kabbalah* (Dordrecht: Kluwer, 1995). Phyllis J. Walton and I argued in "The Geometrical Kabbalahs" that the belief that mathematics is descriptive of reality was in part a contribution to the scientific method from the Kabbalistic-Neoplatonic tradition. Alchemy and the chemical philosophy established weight, measure, experimentation, and corpuscularianism-atomism as part of natural philosophy. These also endure recognizably in modern science. Of course, outside of technical scientific texts, some scientists, such as noted naturalist John Ray (1620–1705), published harmonizing, faith-promoting works. Ray's *Wisdom of God Manifest in the Works of the Creation* (London, 1691) was republished in the nineteenth century. The Boyle lectures, directed to the populace, also retained the connection between science and theology. W. Derham's *Physico-Theology: or a Demonstration of the Being and Attributes of God from His Works of Creation*, the Boyle Lectures of 1711 and 1712, had gone through nine editions by 1737. Still, before the end of the seventeenth century, technical scientfic works were becoming increasingly nontheological. During the eighteenth century, theological references became even rarer.

Epilogue. Creation Science and Intelligent Design: Back to the Future

It may seem strange and even inappropriate to conclude a study of the chemical philosophy of the sixteenth and seventeenth centuries with an essay on its relationship to the interaction of science and religion in the present. Indeed, most historians of early modern natural philosophy work out of interest in the period and its issues with no particular thought of its relevance to specific debates in the late twentieth century and newly arrived twenty-first century. I am and will continue to be one of those historians. Yet, I cannot help but notice that the development of natural philosophy from a religious, Bible-believing environment is related to theological-scientific issues being debated today, largely in the United States of America. From time to time, one reads that Kepler, Galileo, Boyle, or Newton were religious, so why then is science atheistic?[1] In such contexts, the sixteenth and seventeenth centuries are connected to the present. This seems justification enough to express some thoughts on the subject of the relevance of the history of early modern science to the contemporary religion–science debate.

---

1 An interesting discussion of the issue is Steve Paulson's January 2, 2007, interview with Ronald L. Numbers for *Salon* online magazine, http://www.salon.com.

I

Religious belief and acceptance of the Bible were so general in early modern Europe that they influenced all areas of thought. Their shaping of natural philosophy is best seen in specific cases. The universe of discourse (the subject matter deemed appropriate to a discipline) and the method of natural philosophy were greatly entangled with the Bible and religion throughout the sixteenth and seventeenth centuries. Specific natural philosophers took religion very seriously. Chemical philosophers, as shown above, used the first lines of Genesis as a guide to basic matter theory and processes. This fit with the generality of the text and its lack of natural philosophical specifics. Van Helmont, for example, worked out his ideas of water as the basic element and gas with laboratory experiments. Although he and others believed that the soul discovers natural truths in God's creation, they developed a spiritualized empirical experimentation. Over time, the subject matter and method of natural philosophy became more and more divorced from theology.

Nature, as the specific subject matter of natural philosophers, was at the heart of Boyle's disagreement with More. Boyle saw the universe of discourse (he did not use the term) of chemistry as different from that of theology. That is what he meant by discoursing "like a naturalist." Boyle was not alone in his understanding of natural philosophy as being different from theology. Kepler, a devout Christian, rejected the idea that anything but facts properly fit to mathematics is a proper method for astronomy. He wrote to the alchemist Joachim Tanckius:

> I also play with symbols; . . . but when I play I do not forget I am playing. For nothing is proved with symbols, nothing hidden is shown in natural philosophy through geometrical symbols except that known before, unless it is shown by certain reasons to be not only symbolic but also a description of the connection between things and causes.[2]

I believe that the growing awareness of natural philosophy as a separate discipline is a major aspect of the "scientific revolution" of the sixteenth and

---

2 Johannes Kepler, letter to Joachim Tanckius, May 12, 1608, *Gesammelte Werke*, ed. Walther von Dyck, Max Caspar, Franz Hammer, Martha List, and Volker Bialas (Munich: C. H. Beck, 1937–). Translation mine.

seventeenth centuries. Nature was considered a book to be read by specialists who understood its subject matter. Special tools, mathematics, weight, measure, and observation were considered appropriate to interpreting the text. This attitude is apparent in Newton's *Principia* (1687), the supreme accomplishment of seventeenth-century natural philosophy. Newton derived statics, dynamics, and celestial motion from three laws and applied mathematics. He did not invoke hypotheses (causal speculations) or overt theology in the text. The *Principia* is on the face remarkably agnostic. Newton studies have shown him to be deeply religious and well-read in theology. His religious ideas did influence his thought.[3] He, however, like Boyle seems to have regarded natural philosophical discussion as outside theological discourse.

The rejection of Aristotelian first principles and syllogistic method by "new scientists" in the sixteenth century and especially in the seventeenth century did not mean that they did not adhere to certain a priori assumptions. They assumed that nature can be known and that mathematics and measurement are tools that reveal the workings of the world. These assumptions were the essence of a method that was limited to dealing with physical phenomena. The endless debates among divines over theology highlighted a need to foster natural philosophical discussions based on agreed-upon sources of data and rules of interpretation. Focusing on matter and observable phenomena interpreted by a common method led to that end. This, of course, did not mean that natural philosophers rejected theology. They simply excluded it from their technical work. The definition of the subject matter and methodology coupled with the creation of professional organizations such as the Royal Philosophical Society and research institutions such as Greenwich and other observatories turned natural philosophy into science. The term science shifted from a general meaning of study or knowing to a specific one, the study of the natural world.

II

If science became increasingly removed from theology, the reverse was not true. Science's success in explaining nature attracted men of religion, especially in England and later in the United States. Science seemed to reveal the handiwork

---

3 See the biography of Newton, Richard S. Westfall's *Never At Rest* (Cambridge: Cambridge University Press, 1980).

of a supernatural engineer. Theologians, such as More and Cudworth, saw science as a proof for the existence of God. His wisdom was indeed manifest in the Creation (*fig. XII*). The union of science and theology was often called "natural theology." Natural theology was a popular and influential force in the intellectual world until the middle of the nineteenth century when geological fossil analysis and Darwin's studies showed that the alliance between religion and science is fragile.

The crisis between religion and science was probably inevitable given science's eschewal of the theological and concentration on the natural or material world. The outline of the conflict is seen first in the modest shock waves that followed Pierre Simon Laplace's statement to Napoleon that his *System of the World* (1796) had no need for the hypothesis of God.[4] Laplace did not so much draw attention to his doubt of God's existence as he did to the fact that science is bound to naturalistic explanations of the material world. A scientific explanation of the world should have no recourse to the supernatural. As shocking as Laplace's position was to the religious, it did not elicit the level of theological backlash of Darwin's *Origin of Species* (1859) sixty-three years later. The direct relevance of Darwin's biological studies to the Bible's special creations of plants, animals, and man made the *Origin* a more apt vehicle to reveal the discontinuity between religion and science.[5]

Chemistry was not a factor in either Laplace's or Darwin's work. Its transformation from Van Helmont's and Boyle's formulations took place in the eighteenth century and had little to do with biology outside of the discipline of medicine. It can be argued that Antoine Lavoisier's (1743–94) careful researches and formulation of oxygen chemistry constituted a chemical revolution, but it was one of physical not biological chemistry. Lavoisier built on the significant laboratory tradition begun in the seventeenth century and often associated with medicine by Hermann Boerhaave (1668–1738), William Cullen (1710–90), Joseph Black (1728–99), and many others.[6] It is important

4 Bertrand Russell, *Religion and Science* (1995; New York: Oxford University Press, 1961), 58.

5 For a detailed discussion of science and religion in the eighteenth and nineteenth centuries, see John C. Greene, *The Death of Adam* (Ames, IA: Iowa State University Press, 1959).

6 For details, see Debus, *Chemistry and Medical Debate*; Robert G. W. Anderson, "Boerhaave to Black: The Evolution of Chemistry Teaching," *Ambix* 53 (2006): 237–54;

Figure XII. "Adam in Paradice" and title page. John Ray, *The Wisdom of God Manifested in the Works of the Creation.* London, 1762.

and Henry Guerlac, *Lavoisier: The Crucial Year* (Ithaca, NY: Cornell University Press, 1961).

to note that chemistry had little to offer naturalists in the eighteenth and early nineteenth centuries. Naturalists were focused on geology and biology and it was not until the later nineteenth century that chemistry developed the fundamental understanding of atomic and molecular structure that first led to organic chemistry and ultimately to serious biochemistry in the twentieth century.

Darwin was a poor chemist but an exemplary nineteenth-century English naturalist. His impact on the future of natural theology and biblical Creation arose from his observations of fossils, plants, and animal life. His studies allowed him to formulate a theory that species arise naturally over time by means of natural and sexual selection. His theory, like any scientific theory, was useful because of its ability to suggest research topics and make predictions. Darwin's theory predicts, among other things, that later species would not be found in fossil strata with earlier species and that species are continually undergoing changes that will result in the emergence of new species. Darwin's theory ignores natural theology's traditional commitment to the literal truth of Genesis and its scientific corollary of the simultaneous, special creation of fixed species.

In basing his theory solely on the observation of nature, Darwin, an agnostic, does not appear to have consciously wanted to discredit religion. Rather, he simply followed the methodology of science. Like Boyle, he was acting as a naturalist rather than as a theologian. The materialist orientation of Darwin's theory revealed the weakness of natural theology. Because of their different objects of study, science (the natural) and theology (the supernatural) have different universes of discourse.

Even when the pre-nuclear physics of Lord Kelvin (1824–1907) seemed to show that there was insufficient geological time (according to the laws of thermodynamics, the world was about 100,000 years old) for evolution to have taken place, most leading biologists did not seek a return to the pre-Darwin synthesis of natural theology. They believed that Darwin, while not correct, had freed science from the burden of religion. In 1909, William Hanna Thomson, MD, LLD., wrote:

> The chief credit which historically will attach to the name of Charles Darwin is that more than any one else in our times he established the conviction among biologists

> that the processes of life are as completely subject to natural laws as are all other processes in the world.[7]

As this quotation indicates, Darwin's impact on science was, perhaps, as great as his impact on the relationship between science and religion. The debate about biblical literalism that Darwinism engendered made certain aspects and ramifications of the scientific method more apparent. After Darwin, the materialistic orientation of science was undeniable. Science's universe of discourse was ideally seen to be independent of everything except the material world.

While the positivists of the nineteenth century believed that the rigorous study of the material world would lead to truth, it became increasingly clear that scientific explanations could never achieve certainty. The limitations of knowledge as delineated by Hume and those who followed his lead were applied to science. Scientific theories and hypotheses were seen to be both historically and philosophically tentative. There was no demonstrable onward and upward movement toward *the* truth. Rather, there was a succession of facts and theories to account for them. The theories and hypotheses of science were useful devices for placing observations into a framework that allows predictions or technological advancements. They could never be shown to describe reality. One theory became superior to another by virtue of its predictive value, its usefulness rather than its truthfulness. As new theories superseded and incorporated old ones toward the end of the nineteenth century, scientists came to see more clearly the essence of their method. It was not so much a way to discover ultimate truth as it was a means to avoid deception. Predictability and usefulness, which were long both explicit and implicit in science, became more and more the standard for evaluating theories. In the twenty-first century, string theory in physics is being criticized because, among other problems, no one has yet suggested an experiment that might disprove it.[8]

---

7 William Hanna Thomson, *What Is Physical Life?* (New York: Dodd, Mead & Co., 1909), 20–21.

8 See for example, Lee Smolin, *The Trouble with Physics: The Rise of String Theory, the Fall of a Science, and What Comes Next* (New York: Houghton Mifflin, 2006).

III

The fissure between biological science and the Bible occasioned by Darwin was recognized immediately by both sides. Bishop Samuel Wilberforce and Thomas Huxley became dramatic proponents of faith and science. The Bible describes a six-day Creation, and so it was. To Wilberforce, the truth of scripture is absolute. Faith dictates a rejection of any theory that contradicts fundamental biblical facts; such facts should inform science. Huxley could not fathom that the massive data assembled and built into a natural theory by Darwin could be doubted. These first shots fired in 1860 framed a conflict that came to transcend Darwin and go to the independence and integrity of science itself. The debate continued even after organic evolution was seemingly dead as a theory. In 1896, Sir Thistleton Dyer, FRS, wrote to *Nature*, "The Darwinian theory of organic Evolution seems hardly to have a convinced supporter left [in England] except Mr. Alfred R. Wallace and myself!"[9]

Religionists were not the only or even the most strident combatants. The first president of Cornell University, A. D. White, attacked religion as a traditional impediment to scientific inquiry, something not supported by the history of the chemical philosophy. His *A History of the Warfare Between Science and Religion* was a plea to free science from the hegemony of religionists and the Bible. At the same time, many ministers at the end of the nineteenth century defended the Bible from a variety of positions. George D. Armstrong attacked evolution because it conflicted with special creation. He, however, allowed time for geological processes. T. DeWitt Talmage argued that there could be no compromise with science because anything other than biblical literalism removed the miraculous from the world. Religious authorities themselves proposed several explanations of the biblical "days": the six days of creation were six indefinite periods, there were six 24-hour days, there was a gap between the creation of the forms of life found in the fossil record and the creation recorded in Genesis, Genesis is an allegorical account.

The story of creationist theology in the nineteenth and twentieth centuries is well told by Ronald L. Numbers.[10] There is no need here to detail

---

9 T. R. Dyer, Letter, *Nature*, July 30, 1896.

10 Ronald L. Numbers, *The Creationists*, 2nd ed. (1992; New York: Knopf, 2006).

his comprehensive efforts. The majority of denominations eventually accepted science and took a less literal view of the biblical Creation. A devoted minority—many of whom were Adventists—affirmed the literal truth and inerrancy of the Bible. By the end of the first quarter of the twentieth century, Darwinism had been resurrected as *the* general theory for the biological sciences. The discovery of radioactivity provided another source for the earth's heat and altered the understanding of thermodynamics. Celestial time expanded to billions of years. There was sufficient time for the mechanisms of natural and sexual selection to work.

The discovery of Mendelian ratios, chromosomes, and genes provided a physical basis for inheritance and mutation.[11] By the second half of the twentieth century, chemistry possessed tools to analyze genetic matter into its component parts. This led to a race to articulate the structure of DNA, a race won by James Watson, Francis Crick, Maurice Wilkins, and Rosalind Franklin.[12] They showed that DNA is a double helix of known nucleotides.

---

11 August Friedric Leopold Weisman suggested that the mechanism of inheritance was inside the cell in his *Die Kontinuität des Keimplasmas als Grundlage einer Theorie der Vererbung*, 2nd ed. (Jena: Gustav Fischer, 1892). Thomas Hunt Morgan analyzed how pairs of chromosomes carry heredity. He discussed his work in a general lecture in 1922, "Croonian Lecture: On the Mechanism of Heredity," *Proceedings of the Royal Society*, ser. B, vol. 94, no. B659 (December 1, 1922). Morgan received the Nobel Prize in 1933.

Of interest in the context of the chemical philosophy, Peter Severinus discussed Paracelsian pathological seeds, which inhabit the parental seed, as the mechanism of hereditary disease. He wrote in *Idea medicinae*:

> for whatever is contained in the parents, be it healthy or diseased ... is able to affect ... [the] root of human nature.... Thus much is contained in parental seeds which lurk silently, maturing and ready to burst out in time, unexpected symptoms in the astronomy of birth which then become explicable. When this happens ... gouty children [are formed] from nongouty parents and epileptic and mad children from nonepileptic and sane parents. *Vice versa*, gouty parents engender nongouty children, epileptic nonepileptic, parents with the stone, children without the stone. (221; translation mine)

12 The discovery was announced in volumes 171–72 in 1953 in *Nature*: J. D. Watson and F. H. C. Crick, "A Structure for Deoxyribose Nucleic Acid"; M. H. F. Wilkins, A. R. Stokes, and H. R. Wilson, "Molecular Structure of Deoxypentose Nucleic Acids"; R. Franklin and R. G. Gosling, "Molecular Configuration in Sodium Thymonucleate"; J. D. Watson and F. H. C. Crick, "Genetical Implications of the Structure of Deoxyribonucleic Acid"; R. Franklin and R. G. Gosling, "Evidence for 2-Chain Helix in Crystalline Structure of

When researchers compared the DNA of various organisms, they found substantial similarities. This fit the prediction of the theory of organic evolution. Chemistry was in the game as an essential player. Although modern biochemistry no longer shares the theories and social context of the chemical philosophers, their belief that chemistry is the key to biology and effective medicine seems prescient. Nature studied by observation and experiment in the laboratory has indeed yielded many of her fundamental secrets, secrets that can heal.

IV

The revival of organic evolution as the "unified field theory" of biology and the breakthroughs in biochemistry might have relegated literal biblical Creationism to the fundamentalist seminary. However, in 1970, the Institute for Creation Research was established in California. The Institute enlisted credentialed scientists to attack the science underlying the theory of organic evolution. As Numbers notes, the Institute transcended scientific research. It would "prepare creationist curriculum materials for grades one through six ... undertake research on the physical aspects of the flood, and support various 'extension ministries' such as radio programs, seminars and literature."[13] One of the scientists associated with the Institute was Duane Gish, a PhD in biochemistry from the University of California, Berkeley. Gish was an experienced researcher at universities and in private industry. He authored two central texts for the Institute, *Evolution: The Fossils Say No!* (1979) and *Evolution: The Challenge of the Fossil Record* (1985). The Institute's agenda in producing such materials was to promote Creationism as an acceptable theory, competing with Darwinism. Creationists argued that the two theories should at least be taught together. Students could then decide which had the better explanation. Democracy was conisdered a way to evaluate science.

The presence of trained scientists in creation science research (although they often worked outside of their specialties) and the contention of equivalency, if not superiority, of Creationism to Darwinism raises central

---

Sodium Deoxyribonucleate"; and M. H. F. Wilkins, W. E. Seeds, A. R. Stokes, and H. R. Wilson, "Helical Structure of Crystalline Deoxypentose Nucleic Acid."

13 Numbers, *The Creationists*, 284.

questions about the nature of science that seem to return us to the sixteenth and seventeenth centuries. The Aristotelians, typified by the Jesuit Christoph Clavius, saw the study of nature as proscribed by religious doctrine. This was the traditional position. The modernists, such as Kepler and the chemical philosophers, although obviously religious, wanted to follow the book of nature wherever it led them. God was present in the creation and knowing Him was the issue, not theological doctrine. This attitude toward nature was at the root of the dispute between Erastus and the Paracelsians. Erastus elevated theology above experience; that is why he accused the Paracelsians of heresy and using magic.

Today we can formulate the issue thus: Should studies of nature that have been limited by a priori theological dogma be as acceptable as those studies that have not been so limited? This formulation, if answered in the negative, does not exclude the work of scientists who have beliefs that might shape their conceptualizations. The history of science is replete with believers, agnostics, and atheists who have done both good and bad science. If we construe science as a method that allows theories to be tested (Can data disprove the theory?), then the theories and work of anyone can be evaluated by the rules of the discipline. The problem arises when the discipline rejects research or theories and the rejection is not accepted. Creation scientists, to one degree or another, will not take no for an answer. Their commitment to literalism ultimately trumps the rules of science. Their reaction is not unlike the attitude of theologians of the sixteenth and seventeenth centuries who, based on a priori assumptions, rejected aspects of the "new science." Erastus could not overlook the Paracelsian heresy. Van Helmont died under "house arrest" of the Roman Catholic Church. I have not discussed Galileo in this work, but he also died under ecclesiastical house arrest because he treated Copernicanism as more than a hypothesis after he had received a stern ecclesiastical warning not to do so.[14]

Any belief that shapes scientific research can be asked the question, "What does this belief add to the research?" In the case of the chemical philosophy, the Genesis creation account was crucial to forming basic ideas and an epistemology. Modern biblical literalism, however, allows fewer

---

14 See Giorgio de Santillana, *The Crime of Galileo* (Chicago: University of Chicago Press, 1955). Galileo's conflict with the church has created a vast literature, but this is the best place to start.

possibilities than the several interpretive approaches available in the sixteenth and seventeenth centuries, which are outlined in chapter 1. It is possible that a belief in special creation could help a scientist to formulate research questions. The questions are not the issue. The laboratory work and reporting of data are. Can Creationism, and the Creationists, accept data that contradicts the answer expected by belief? Van Helmont and Boyle reported their data and refined their thought accordingly. Boyle disagreed with More about whether there should be theological constraints to natural thought. Boyle's attitude has become the sine qua non of science from the eighteenth century to the present. The tentative nature of science requires a certain skepticism that is difficult for true believers of any stripe to maintain. The result of engaging the data with a skeptical mind, however, is likely to be abandonment of "true" belief. Numbers, for example, tells us that several "young Earth" Creationists have rejected their former beliefs after studying geology.[15]

The fact that science's limited universe of discourse excludes God as an explanatory factor differentiates it from biblical literalism. Many people who are troubled by science's agnosticism would like it to acknowledge God's existence as the intelligent designer of the universe. Some scientists and philosophers believe that their agnostic universe of discourse is the only relevant one; therefore, as a result of their disciplines' agnosticism, they become atheists or agnostics.[16] The disbelief in God can become an axiom in science courses. The Intelligent Design movement is, at least in part, a reaction to scientific agnosticism. The Discovery Institute of Seattle, Washington, is an example of an organization of people who realize that they cannot keep Darwin out of the schools. Now they argue that the concept of an intelligent designer of the universe should be offered to students in contrast to what is perceived to be scientific atheism.

The question of intelligent design is clearly related to the rise of science in the early modern era. In the sixteenth and seventeenth centuries, virtually all natural philosophers believed that God was the creator of an orderly, intelligible world. In fact, that belief led to the crucial a priori assumption that man can understand nature. The assumption helped establish mathematics, which like God is ideal and unchanging, as a tool for reading the book of

---

15 Numbers, *The Creationists*, 290–96.

16 Numbers believes that the majority of scientists are theists, *Salon* interview.

nature. The chemical philosophers' epistemology was rooted in God's continuing presence in his creation. There is no doubt that intelligent design was a given in science for a significant period and was built into its method.

"God talk" in science declined as the discussion of Boyle and Newton in the previous chapter demonstrated. Science's exclusion of God as an explanatory factor in favor of physical, testable facts created a shared universe of discourse among scientists. What acknowledging intelligent design would add to scientific research and theorizing is not clear. It certainly would make some theists feel better. A discussion of intelligent design seems more appropriate in a philosophical context. Science deals with proximal, transitory explanations and their evaluation. Intelligent design is an ultimate issue. The Bible may state that the heavens declare God's glory, but the scientist is limited to questions of the nature of light or gravity, mere artifacts of God's presence. The beauty and order of nature prized by some can be seen as frightening chaos by others. The return of theism in the form of intelligent design seems to ignore science's limited scope. It would be an anachronistic invocation of sixteenth- and seventeenth-century thought in the twenty-first century.

V

Modern Creationism goes beyond the philosophical concept of intelligent design when it challenges science for not using the truth of Genesis to interpret nature. It is critical of science while at the same time accepting that Genesis is the absolute word of God. Genesis is a literal description of the creation and cannot be questioned. This is a much more restricted view than that held by biblical literalists in the past. As chapter 1 demonstrates, the Bible's poetic ambiguity allows a variety of interpretive approaches and readings even among true believers in its divine origin. The most significant threat to present-day biblical literalism is not the tentative proximal explanations of science but rather biblical criticism, as I noted in the preface.

Modern biblical criticism grew out of the scientific worldview. It became well established in the nineteenth century. Scholars came to see the Bible as a text to be studied as they studied other texts, not as an inviolable revelation. Although for centuries inconsistencies in the Bible's text had been apparent to biblical commentators, they employed various traditional approaches to explain them. By the seventeenth century, some individuals, such as Baruch Spinoza, had come to believe that the Bible was part of a progressive revelation, composed by inspired humans.[17] Even believers acknowledge virtually insurmountable difficulties in knowing exactly what such an ancient text means. Robert Boyle was one of those. As a Hebraist and Bible scholar he wrote,

> [F]or I consider in the second place, that not only we have lost divers of the significations of many of the Hebrew words and phrases, but that we have also lost the means of acquainting ourselves with a multitude of particulars relating to the topography, history, rites, opinions, fashions, customs, &c. of the antient Jews and neighbouring nations, without the knowledge of which we cannot, in the perusing of books of such antiquity, as those of the Old Testament, and written by and (principally) for Jews; we cannot, I say, but lose very much.[18]

Moshe H. Goshen-Gottstein has discussed the early development of critical biblical studies.[19] The story is somewhat convoluted, but fascinating. It can be telescoped for the purpose of this essay to the nineteenth century. Like nineteenth-century "scientific historians," biblical scholars, especially Germans, narrowed the universe of discourse for their discipline by excluding God and revelation as explanatory factors. They studied the text, and compared it with other texts and with archaeology in order to sift through editorial debris and arrive at the elements from which the Bible was composed. These scholars constructed a general theory of biblical development that stressed text units, editing, and evolution. Julius Wellhausen's (1844–1918) restatement of K. H. Graf's (1815–69) work became known as the Graf-Wellhausen hypothesis. It functioned, and still functions, as the biblical equivalent of Darwin's theory.[20]

---

17 Baruch Spinoza, *Tractatus Theologico-Politicus* (1670). See Williams, *Common Expositor*, 255–68, for a discussion of the beginnings of this critical approach.

18 Boyle, *Some Considerations Touching the Style of the Holy Scriptures*, in *Works*, 2:259.

19 Goshen-Gottstein, "Foundations of Biblical Philology in the Seventeenth Century."

20 Julius Wellhausen, *Die Composition des Hexateuchs* (Berlin, 1885) originally published as articles in *Jahrbücher für Deutsche Theologie*, 21 (1876): 392–602, and 22 (1877): 467–79. For a discussion of the development of modern biblical studies, traditional interpretation, and their relationship to religious fundamentalism, see James L. Kugel, *How to Read the Bible: A Guide to Scripture Then and Now* (New York: Free Press, 2007).

While no longer accepted in many particulars, no subsequent theory has achieved consensus. Wellhausen stated that the Pentateuch was not written by Moses. It consists of four basic textual traditions, J, E, P, and D, that had been edited over time and finally collected and intertwined in the five books of the traditional Pentateuch and the Book of Joshua. Each textual tradition is marked by thematic and vocabulary distinctions, such as the use of YHWH in J and Elohim in E.

Such a critical approach to the Bible, if correct or even partially correct, undermines the claim that Genesis is absolutely true; moreover, it disconnects Genesis from science, in particular as the a priori starting point of scientific knowledge. How critical scholars analyze Genesis illustrates both the force of their method and its challenge to Creationism. Among many issues that interest scholars are anachronisms, duplicated stories, and obvious textual interweaving. Genesis 36:31–40 lists a series of Edomite kings who ruled "before any king reigned over the Israelites." This is obviously written at some time after Saul became the first king of Israel, not hundreds of years before, during the time of Moses. Other examples are references to the land of Dan (Gen. 14:14) and Philistines (Gen. 21:34, 26:14, 15, and 18) before their time.

Genesis also contains different versions of the same stories. Genesis 21:31 attributes the name Beersheba to a covenant between Abraham and Abimelech, but Genesis 26:31 associates the name with Isaac. Abraham pretends that Sarah is his sister in order to deceive Pharaoh (Gen. 12:10–20) and then Abimelech (Gen. 20:1–18), a ruse used later by Isaac when Abimelech wants Isaac's wife Rebekah (Gen. 26:6–11). There are two stories changing Jacob's name to Israel (Gen. 32:2 and Gen. 35:10). Scholars believe that these duplications show that the editor(s) of Genesis drew on the stories of different, but related, traditions.

Of special significance for creation science and the biblical text are the conflicting passages about Noah's flood and the presence of the two creation stories in the first chapters of Genesis. Looking at the Flood story first, God commands Noah to take a pair of each species into the ark in Genesis 6:19, but seven pairs of clean and one pair of unclean animals in Genesis 7:2. Yet, verse 8 refers to only one pair of each entering. The length of the Flood is similarly uncertain (compare Genesis 8:6 ff. with Genesis 7:24).

The creation stories of Genesis 1 and 2 differ greatly. Genesis 1:1 through 2:4 gives a poetic description of a transcendent God (Elohim) making the world through the power of his word. He says, "let us make man" and makes "male and female." The rabbis taught that this "man" was one being, an androgynous creature. The second account is more prosaic. YHWH-Elohim creates plants, man from the dust, and a garden for man to tend. Later, a woman is made from the man. The man and woman violate God's command and hide from him, as he moves through the garden. He finds them, expels them from the garden, and pronounces their punishment. Language, theme, and the nature of God differ in the two stories. Scholars conclude that the two texts represent different traditions, put together to form one creation myth.

Biblical scholars are not alone in their view of the Bible as an edited work. Archaeologists, whose mindset differs considerably from text-critical readers, have found serious discrepancies between the biblical story and the evidence from excavations. They find only ambiguous or contradictory evidence for the stories of the patriarchs. The details of Abraham's life are very general and could apply to nomads in many periods. But the camel, which is included in the Abraham story, was not domesticated until about 1000 BCE, approximately 800 years after Abraham should have existed. The import of archaeology is that the Bible, especially the book of Genesis, is not reliable as to historical facts and context.[21] Both readers (textual critics) and diggers (archaeologists) apply the scientific method with its limited universe of discourse to the Bible with negative results for literalists.

CONCLUSION

Why science developed from the religio-social context in sixteenth- and seventeenth-century Europe is a complicated question. Its rise at that time, however, insured that Christianity and the Bible would influence it. The subject of this book is a case study of Genesis and the chemical philosophy. Over years of research, I became aware of the significant role that Genesis had played in the thought of Paracelsus and his heirs. At a time when everyone believed that the Bible was a divinely revealed text, it is unremarkable that it would influence natural philosophers. This is especially true given that there was no sense of the vast expanse of time since the creation moment. The specific ways that natural

---

21 Several archaeologists have written popular descriptions of their work. See, Israel Finkelstein and Neil Asher Silberman, *The Bible Unearthed* (New York: Free Press, 2001), and William G. Dever, *What Did the Biblical Writers Know and When Did They Know It?* (Grand Rapids, MI: Eerdmans, 2001).

philosophers used the Bible are important to our understanding of how science arose. The intellectual and social factors of Europe in the sixteenth and seventeenth centuries that gave rise to modern science can never be duplicated, but we can examine and attempt to understand the importance of religious thought to natural philosophy in that era.

To say that Paracelsus, Kepler, Van Helmont, Boyle, and Newton were religious is meaningful in their historical context. It is not very meaningful to modern science and its method or to scientific Bible study. How the founders of the new science would have reacted to modern science, or, for that matter, to biblical literalism or liberal religion is absolutely unknowable. I have speculated on some points of convergence in defining the limits of natural philosophy, method and research goals that tentatively associate sixteenth- and seventeenth-century natural thought with what science has become. Such thought experiments are enjoyable, but they do not add much to contemporary debates on science and religion. Nor should one twist our understanding of the scientific revolution to fit modern concerns. Science and theology have diverged so significantly that they do not now share the same universe of discourse. Today, with regard to science, we see the importance of Boyle's insight; scientists speak as naturalists and not as theologians.

SELECT BIBLIOGRAPHY

Primary Sources

Ashmole, Elias. *Theatrum Chemicum Britannicum*. London, 1652.

Bacon, Francis. *Sylva Silvarum or a Natural History in Ten Centuries*. London, 1627.

Basil Saint, Bishop of Caesarea. *St. Basil: Letters and Select Works*. Trans. with notes by Blomfield Jackson, vol. 8 in *A Select Library of Nicene and Post-Nicene Fathers*. 2nd ser. New York: Christian Literature, 1895. (Alternate title: *The Treatise de Spiritu Sancto, The Nine Homilies of the Hexaemeron and the Letters of Saint Basil the Great*) http://www.ccel.org/ccel/schaff/npnf208.html.

Becher, Johann Joachim. *Psychosophia oder Seelen-weissheit*. Frankfurt, 1678.

Boyle, Robert. *Some Considerations Touching the Style of the Holy Scriptures* in *The Works of the Honorable Robert Boyle*. Ed. Thomas Birch. vol. 2. London, 1772. 6 vols.

———. *A Hydrostatical Discourse occasion'd by some Objections of Dr. Henry More*. London, 1672.

———. *The Sceptical Chymist*. London, 1661; repr., London: Dent, 1911.

Calvin, John. *Commentary on Genesis*. Trans. John King. Edinburgh: Calvin Translation Society, 1847.

The Chaldaean Oracles as Set Down by Julianus, Translated into Latin by Francesco Patrizzi andTranslated into English by Thomas Stanley.

Berkeley Heights, NJ: Heptangle Books, 1989. Originally published as *The Chalderick Oracles of Zoroaster and his Followers*. . . . Ed. and trans. Thomas Stanley. London, 1661. http://www.esotericarchives.com/oracle/oraclesj.htm.

Charleton, Walter. *Physiologia Epicuro-Gassedno-Charltoniana*. London, 1654; repr., New York: Johnson Reprints, 1966.

Cicero. *De Natura Deorum*. Leipzig, 1818.

Clavius, Christoph. *In Sphaeram Ioannis de Sacro Bosco Commentarius*. Venice, 1596.

Croll, Oswald. *Basilica Chymica* . . . published with *Tractatus novus de Signaturis rerum internis*. Frankfurt, 1611.

Cudworth, Ralph. *The True Intellectual System of the Universe, with References, and the Author's Life by Thomas Birch*. New York: Andover, 1837; originally published London, 1678.

De Lyra, Nicholas. *Postilla super totam Bibliam*. Nuremberg, 1481.

Eusebius of Caesarea [Eusebius Pamphili]. *Ecclesiastical History*. Trans. Isaac Boyle. Grand Rapids, MI: Baker Book House, 1966.

Ficino, Marsilio. *Omnia Divini Platonis Opera*. Venice: Iannem Mariam Bonellum, 1556.

Fludd, Robert. *Mosaicall Philosophy*. London, 1659.

———. *Utriusque Cosmi, Tomi Secundi*. Oppenheim, 1619.

Glauber, Johann Rudolph. *The Works*. Trans. Christopher Packe. London, 1689.

Grosseteste, Robert. *On the Six Days of Creation*. Trans. C. F. S. Martin. Oxford: British Academy by Oxford University Press, 1996.

Janitor Pansophus. In *Musaeum Hermeticum*. Frankfurt: L. Jennisii, 1625. And in *Musaeum Hermeticum Reformatium et Amplificatum*. Frankfurt: Hermannum á Sande, 1677.

Kepler, Johannes. *Gesammelte Werke*. Ed. Walther von Dyck, Max Caspar, Franz Hammer, Martha List, and Volker Bialas. Munich: C. H. Beck, 1937–.

Khunrath, Heinrich. *Amphitheatrum sapientiae aeternae*. Hamburg, 1595.

———. *Confessio de chao*. English translation. British Library. Ashmole MS 1459.

———. *Confessio de chao physico-chemicorum catholico*. Magdeburg, 1596.

———. *De igne magorem philosophorum*. Strassburg, 1608.

———. *Symbolum physico-chymicum*. Hamburg, 1598.

———. *Von hylealischen das ist, pri-materialischen catholischen, oder algemeinemy natürlichen . . . chaos*. Magdeburg, 1597.

Libavius, Andreas. *Alchemia*. Frankfurt, 1597.

———. *Appendix Necessaria Syntagmatis Arcanorum Chymicorum*. Frankfurt, 1615.

———. *De Universitate et Originibus Rerum Conditarum Contemplatio Singularis Theologica, et Philosophica, Iuxta Historiam Hexaëmeri Mosaici. . . .* Frankfurt, 1610.

Linden, Stanton J., ed. *William Cooper's A Catalogue of Chymicall Books, 1673–88*. New York: Garland, 1987.

Lucretius. *De Rerum Natura*. Cambridge, 1686.

———. *Of the Nature of Things*. Trans. William E. Leonard. London: E. P. Dutton, 1921.

Luther, Martin. *Lectures on Genesis*. In vol. 1 of *Luther's Works*. Ed. Jaroslav Pelikan. St. Louis, MO: Concordia, 1958.

Maimonides, Moses. *The Guide of the Perplexed*. Trans. Shlomo Pines. Chicago: University of Chicago Press, 1963.

Marinus. *Life of Proclus*. Trans. Kenneth Sylvan Guthrie. Grand Rapids, MI: Phanes Press, 1986.

Mersenne, Marin. *Quaestiones caleberrimae in Genesim*. Paris, 1623.

Midrash Rabbah, Genesis. Trans. H. Freedman. London: Soncino Press, 1983.

Mikraoth Gedoloth. Jerusalem, Isr.: B. Bruchman, 1995.

More, Henry. *Conjectura Cabbalistica*. London, 1653.

———. *The Immortality of the Soul* in *A Collection of Several Philosophical Writings of Dr. Henry More*. London, 1662.

Origen. *Homilies on Genesis and Exodus*. In *Fathers of the Church: Origen*. Trans. Robert E. Heine. Washington DC: Catholic University Press, 1982.

Origins: Creation Texts from the Ancient Mediterranean. Ed. and trans. Charles Doria and Harris Lenowitz. Garden City, NJ: Anchor Books, 1976.

Paracelsus [Theophrastus Bombast von Hohenheim]. *The Hermetic and Alchemical Writings of Paracelsus.* 2 vols. Trans. Arthur E. Waite. Berkeley, CA: Shambhala, 1976; originally published, London: James Elliot, 1894.

———. *Theophrast von Hohenheim genannt Paracelsus: Sämtliche Werke*. Ed. Karl Sudhoff. 14 vols. Berlin and Munich: Oldenbourg, 1923–33.

Philo. *Works.* Trans. William Heinemann. 12 vols. Cambridge, MA: Harvard University Press, 1929–53.

Pico Della Mirandola, Giovanni. *On the Dignity of Man, On Being and the One, Heptaplus*. Trans. Douglas Carmichael. New York: Bobbs-Merrill, 1965.

Pinnell, H, ed. and trans. *Philosophy Reformed and Improved, in Four Profound Tractates.* London, 1657.

Ptolemy, Claudius. *Almagest.* Trans. R. Catesby Taliaferro. In vol. 16 of *Great Books of the Western World*. Chicago: Encyclopedia Britannica, 1952.

Quercetanus, Josephus. [Joseph Duchesne]. *The Practice of Chymicall and Hermeticall Physicke.* Trans. Thomas Timme. London, 1605.

Ray, John. *The Wisdom of God Manifested in the Works of Creation.* London, 1691.

Reuchlin, Johannes. *De Arte Cabalistica.* Hagenau, 1517.

Robinson, James, ed. *The Nag Hammadi Library.* San Francisco: Harper and Row, 1977.

Sandivogius, Micheel [Michael Sendivogius]. *New Light of Alchymy*. Trans. J[ohn] F[rench], M.D. London, 1674.

Severinus, Peter. *Idea medicinae philosophicae.* Basel, 1571.

Steuco, Agostino. *Cosmopaeia*. Venice, 1535.

———. *Recognitio Veteris Testamenti ad Hebraicam Veritatem*. Venice, 1529.

The Sibylline Oracles. Trans. Milton S. Terry. New York: Eaton and Mains, [1899]. http://www.sacred-texts.com/cla/sib/sib.pdf.

Talmud, Babylonian [Talmud Bavli]. Ed. R. Dr. I. Epstein. New York: Traditional Press, n.d.

Theatrum Chemicum. 6 vols. Strasbourg: Zetzner, 1659.

Thomson, William Hanna. *What Is Physical Life?* New York: Dodd, Mead & Co., 1909.

Trismosin, Salomon. *Splendor Solis*. (1582). Trans J. K. London: Keagan Paul, Trench, Truber, and Co., 1921.

Tymme, Thomas. *A Dialogue Philosophicall*. London, 1612.

Van Helmont, Joan Baptista. *Aufgang der Artzney-Kunst*. Sulzbach, 1683.

———. *Oriatrike*. Trans. J[ohn] C[handler]. London, 1662.

———. *Ortus Medicinae*. Amsterdam, 1648; 3rd ed., Amsterdam, 1652.

Von Rosenroth, Christian Knorr. *Kabbalah Denudata*. Sulzbach, 1677.

Weigel, Valentin. *Astrology Theologised*. London, 1649; repr., London: Dryden Press, 1886.

Secondary Sources: Books

Benz, Ernst. *Christian Kabbalah*. Trans. K. W. Wesche. St. Paul, MN: Grailstone Press, 2004.

Brooke, John Hedley. *Science and Religion*. Cambridge: Cambridge University Press, 1991.

Clagett, Marshall. *Greek Science in Antiquity*. New York: Collier Books, 1955.

Clericuzio, Antonio. *Elements, Principles and Corpuscles: A Study of Atomism and Chemistry in the Seventeenth Century*. Dordrecht: Kluwer, 2000.

Cohn, Norman. *The Pursuit of the Millennium*. New York: Essential Books, 1957.

Copenhaver, Brian P. *Hermetica*. Cambridge: Cambridge University Press, 1992.

Coudert, Allison P. *The Impact of the Kabbalah in the Seventeenth Century*. Leiden: E. J. Brill, 1999.

———. *Leibnitz and the Kabbalah*. Dordrecht: Kluwer, 1995.

Crombie, A. C. *Medieval and Early Modern Science*. 2 vols. New York: Doubleday Anchor Books, 1959.

———. *Robert Grosseteste and the Origins of Experimental Science*. Oxford: Clarendon Press, 1953.

Cross, Frank Moore, and Shemaryahu Talmon, eds. *Qumran and the History of the Biblical Text*. Cambridge, MA: Harvard University Press, 1975.

Debus, Allen G. *Alchemy and Early Modern Chemistry, Papers from "Ambix."* London: Society for the History of Alchemy and Chemistry, 2004.

——— *The Chemical Philosophy*. New York: Science History Publications, 1977.

———. *Chemistry and Medical Debate: Van Helmont to Boerhaave*. Canton, MA: Watson Publishing, 2001.

———. *The English Paracelsians*. New York: Franklin Watts, 1966.

———. *The French Paracelsians*. Cambridge: Cambridge University Press, 1991.

———. *Robert Fludd and His Philosophical Key*. New York: Science History Publications, 1979.

———. *Science and Education in the Seventeenth Century: The Webster-Ward Debate*. London: MacDonald, 1970.

Debus, Allen G., and Robert P. Multhauf. *Alchemy and Chemistry in the Seventeenth Century*. Los Angeles: William Andrews Clark Memorial Library, 1966.

Debus, Allen G., and Michael T. Walton, eds. *Reading the Book of Nature*. Kirksville, MO: Sixteenth Century Journal Publishers, 1998.

De Santillana, Giorgio. *The Crime of Galileo*. Chicago: University of Chicago Press, 1955.

Dever, William G. *What Did the Biblical Writers Know and When Did They Know It?* Grand Rapids, MI: Eerdmans, 2001.

Dobbs, Betty Jo Teeter. *The Foundations of Newton's Alchemy or "The Hunting of the Green Lyon."* Cambridge: Cambridge University Press, 1977.

Ferguson, Everett. *Encyclopedia of Early Christianity*. New York: Garland, 1990.

Finkelstein, Israel, and Neil Asher Silberman. *The Bible Unearthed*. New York: Free Press, 2001.

Gillespie, Charles Coulston. *Genesis and Geology*. Cambridge, MA: Harvard University Press, 1951.

Godwin, Joscelyn. *Robert Fludd, Hermetic Philosopher and Surveyor of Two Worlds*. London: Thames and Hudson, 1979.

Goldammer, Kurt. *Paracelsus in Neuen Horizonten: Gesammelte Aufsätze*. Wien: Verband der Wissenschaftlichen Gesallschaften Österreichs, Verlag, 1986.

Goldish, Matt. *Judaism in the Theology of Isaac Newton*. Dordrecht: Kluwer, 1998.

Grant, Edward. *In Defense of the Earth's Centrality and Immobility*. Transactions of the American Philosophical Society. vol. 74, pt. 4. Philadelphia: American Philosophical Society, 1989.

Greene, John C. *The Death of Adam*. Ames, IA: Iowa State University Press, 1959.

Guerlac, Henry. *Lavoisier: The Crucial Year*. Ithaca, NY: Cornell University Press, 1961.

Hannaway, Owen. *The Chemists and the Word: The Didactic Origins of Chemistry*. Baltimore: Johns Hopkins University Press, 1975.

Henry, John. *Knowledge Is Power: How Magic, the Government and an Apocalyptic Vision Inspired Francis Bacon to Create Modern Science*. Cambridge, Eng.: Icon Books, 2002.

Hillerbrand, Hans J., ed. *Radical Tendencies in the Reformation*. Kirksville, MO: Sixteenth Century Journal Publishers, 1988.

Holmyard, E. J. *Alchemy*. Harmondsworth, Middlesex, UK: Penguin Books, 1957.

Hunter, Michael. *Robert Boyle, 1627–91: Scrupulosity and Science*. Woodbridge, UK: Boydell Press, 2000.

Koyré, Alexandre. *Mystique, spirituels, alchemistes du xvi siècle allemand*. Paris: Armand Colin, 1955.

Kugel, James L. *The Bible As It Was*. Cambridge, MA: Harvard University Press, 2000.

———. *How to Read the Bible: A Guide to Scripture Then and Now*. New York: Free Press, 2007.

Kugel, James L., and Rowen A. Greer. *Early Biblical Interpretation*. Philadelphia: Westminster Press, 1986.

Kühlmann, Wilhelm, and Joachim Telle, eds. *Der Frühparacelsismus*. Tübingen: Niemeyer, 2001.

Levine, Lee I. *Caesarea Under Roman Rule*. Leiden: E. J. Brill, 1975.

Lieb, Michael. *The Dialectics of Creation*. Boston, MA: University of Massachusetts Press, 1970.

Lieberman, Saul. *Greek in Jewish Palestine*. New York: Jewish Theological Seminary, 1942.

Linden, Stanton J., ed. *The Alchemy Reader: From Hermes Trismegistus to Isaac Newton*. Cambridge: Cambridge University Press, 2003.

———, ed. *Mystical Metal of Gold*. Brooklyn, NY: AMS Press, 2007.

Lüthy, Christoph, John E. Murdoch, and William R. Newman, eds. *Late Medieval and Early Modern Corpuscular Matter Theories*. Leiden: E. J. Brill, 2001.

Margolin, Jean-Claude, and Sylvain Matton, eds. *Alchimie et philosophie à la renaissance*. Paris: J. Vrin, 1993.

McEvoy, James. *The Philosophy of Robert Grosseteste*. Oxford: Clarendon Press, 1982.

McKnight, Stephen A. *The Religious Foundations of Francis Bacon's Thought*. Columbia, MO: University of Missouri Press, 2006.

Moran, Bruce T. *Andreas Libavius and the Transformation of Alchemy*. Sagamore Beach, MA: Science History Publishers, 2007.

———. *Distilling Knowledge: Alchemy, Chemistry and the Scientific Revolution*. Cambridge, MA: Harvard University Press, 2005.

Morgenbesser, Sidney, et al., eds. *Philosophy, Science and Method*. New York: St. Martin's Press, 1969.

Neuser, Wilhelm H., and Brian G. Armstrong, eds. *Calvinus Sincerioris Religionis Vindex*, Sixteenth Century Essays and Studies, vol. 36. Kirksville, MO: Sixteenth Century Journal Publishers, 1997.

Newman, William R. *Promethean Ambitions: Alchemy and the Quest to Perfect Nature*. Chicago: University of Chicago Press, 2004.

———. *The Summa perfectione of Pseudo-Geber*. Leiden: E. J. Brill, 1991.

Newman, William R., and Lawrence M. Principe. *Alchemy Tried in the Fire: Starkey, Boyle and the Fate of Helmontian Chemistry*. Chicago: University of Chicago Press, 2002.

Nock, A. D. *Early Gentile Christianity and Its Hellenistic Background*. New York: Harper, 1964.

Numbers, Ronald L. *The Creationists*, 2nd ed. 1992; New York: Knopf, 2006.

Pagel, Walter. *Joan Baptista Van Helmont*. Cambridge: Cambridge University Press, 1982.

Parker, James. *The Conflict of the Church and the Synagogue*. New York: Atheneum, 1979.

Partington, J. R. *A History of Chemistry*. 4 vols. London: Macmillan, 1961.

Perella, Nicolas J. *The Kiss Sacred and Profane*. Berkeley: University of California Press, 1969.

Principe, Lawrence M. *The Aspiring Adept: Robert Boyle and His Alchemical Quest*. Princeton, NJ: Princeton University Press, 1998.

Read, John. *Prelude to Chemistry*. 1936; Boston, MA MIT Press, 1960.

Rossi, Paolo. *Francis Bacon: From Magic to Science* Chicago: University of Chicago Press, 1968 [originally published in Italian, 1937].

Runia, David T. *Exegesis and Philosophy: Studies on Philo of Alexandria*. Aldershot: Variorum, 1990.

Russell, Bertrand. *Religion and Science*. 1935; New York: Oxford University Press, 1961.

Sanders, E. P. *Paul and Palestinian Judaism*. Philadelphia: Fortress Press, 1977.

Scholer, David M., ed. *The Works of Philo*. Peabody, MA: Hendrickson, 2002.

Shackelford, Jole. *A Philosophical Path for Paracelsian Medicine*. Copenhagen: Museum Tusculanum Press, 2004.

Shumaker, Wayne. *The Occult Sciences in the Renaissance*. Berkeley, CA: University of California Press, 1972.

Smalley, Beryl. *The Study of the Bible in the Middle Ages*. Notre Dame, IN: University of Notre Dame Press, 1964.

Smith, P. H. *The Business of Alchemy: Science and Culture in the Holy Roman Empire*. Princeton, NJ: Princeton University Press, 1994.

Smolin, Lee. *The Trouble with Physics: The Rise of String Theory, the Fall of a Science, and What Comes Next*. New York: Houghton Mifflin, 2006.

Southern, R. W. *Robert Grosseteste: The Growth of an English Mind in Medieval Europe*. Oxford: Clarendon Press, 1986.

Steneck, Nicholas H. *Science and Creation in the Middle Ages*. Notre Dame, IN: University of Notre Dame Press, 1976.

Szulakowska, Urszula. *The Alchemy of Light*. Leiden: E. J. Brill, 2000.

Theerman, P. H., and K. Hunger Parshall, eds. *Experiencing Nature*. Dordrecht: Kluwer, 1997.

Walker, D. P. *Spiritual and Demonic Magic from Ficino to Campanella*. London: University of London, 1958.

Weeks, Andrew. *Valentin Weigel (1533–1588)*. Albany, NY: State University of New York Press, 2000.

Wellhausen, Julius. *Die Composition des Hexateuchs*. Berlin, 1885.

Westfall, Richard S. *Never At Rest*. Cambridge: Cambridge University Press, 1980.

White, A. D. *A History of the Warfare Between Science and Religion*. New York: Appleton, 1896.

White, Ralph, ed. *The Rosicrucian Enlightenment Revisited*. Hudson, NY: Lindisfarne Books, 1999.

Williams, Arnold. *The Common Expositor*. Chapel Hill: University of North Carolina Press, 1948.

Williams, Gerhild Scholz, and Charles D. Gunnoe Jr., eds. *Paracelsian Moments*. Kirksville, MO: Truman State University Press, 2002.

Wind, Edgar. *Pagan Mysteries in the Renaissance*. New Haven, CT: Yale University Press, 1958.

Wirszubski, Chaim. *Pico della Mirandola's Encounter with Jewish Mysticism*. Cambridge, MA: Harvard University Press, 1989.

Yates, Frances A. *The Rosicrucian Enlightenment*. 1972; London: Routledge Classics, 2002.

Secondary Sources: Articles

Anderson, Robert G. W. "Boerhaave to Black: The Evolution of Chemistry Teaching." *Ambix* 53 (2006): 237–54.

Burnett, D. Graham. "The Cosmogonic Experiments of Robert Fludd: A Translation with Introduction and Commentary." *Ambix* 46 (1999): 113–70.

Crowther-Heyck, Kathleen. "Wonderful Secrets of Nature: Natural Knowledge and Religious Piety in Reformation Germany." *Isis* 94 (2003): 253–73.

Daniel, Dane T. "Invisible Wombs: Rethinking Paracelsus's Concept of Body and Matter." *Ambix* 53 (2006): 129–42.

Debus, Allen G. "An Elizabethan History of Medical Chemistry." *Annals of Science* 18 (1962): 1–29.

———. "Fire Analysis and the Elements in the Sixteenth and Seventeenth Centuries." *Annals of Science* 23 (1967): 127–47.

Forshaw, Peter J. "The Early Alchemical Reception of John Dee's *Monas Hieroglyphica*." *Ambix* 52 (2005): 247–69.

Harley, David. "Rychard Bostock of Tandridge, Surrey (ca. 1530–1605), M.P., Paracelsian Propagandist and Friend of John Dee." *Ambix* 47 (2000): 29–36.

Janacek, Bruce. "Thomas Tymme and Natural Philosophy: Prophecy, Alchemical Theology and the Book of Nature." *Sixteenth Century Journal* 30 (1999): 987–1007.

Linden, Stanton J. "Alchemy and Eschatology in Seventeenth Century Poetry." *Ambix* 31 (1984): 102–24.

McGuire, J. E., and P. M. Rattansi. "Newton and the Pipes of Pan." *Notes and Records of the Royal Society of London* 21 (December 1966): 108–43.

Montgomery, John Warwick. "Cross, Constellation and Crucible." *Ambix* 11 (1963): 65–86.

Moran, Bruce T. "Paracelsus, Religion, and Dissent: The Case of Philipp Homagius and Georg Zimmermann." *Ambix* 43 (1996): 65–79.

Nauert, Charles G., Jr. "The Clash of Humanists and Scholastics: An Approach to Pre-Reformation Controversies." *Sixteenth Century Journal* 4 (1973): 5–18.

Pagel, Walter. "Paracelsus and the Neoplatonic and Gnostic Tradition." *Ambix* 8 (1960): 125–66.

Sailor, Danton B. "Moses and Atomism." *Journal of the History of Ideas* 25 (1964): 3–16.

Snobelen, Steven D. "Isaac Newton, Heretic: The Strategies of a Nicodemite." *British Journal of the History of Science* 32 (1999): 381–419.

Szulakowska, Urszula. "John Dee and European Alchemy." The Durham Thomas Harriot Seminar Occasional Paper, no. 21. Durham, Eng: University of Durham, 1996..

Taylor, F. Sherwood. "The Origins of Greek Alchemy." *Ambix* 1 (1937): 30–47.

———. "Zosimos, *The Visions of Zosimos*." *Ambix* 1 (1937): 88–92.

Thompson, R. Campbell. "A Survey of the Chemistry of Assyria in the Seventh Century B.C." *Ambix* 2 (1938): 3–16.

Waddington, Raymond B. "The Sun at the Center: Structure as Meaning in Pico della Mirandola's *Heptaplus*." *Journal of Medieval and Renaissance Studies* 3 (1973): 69–86.

Walton, Michael T. "Boyle and Newton on the Transmutation of Water and Air, From the Root of Helmont's Tree." *Ambix* 27 (1980): 11–18.

Walton, Michael T., Robert M. Fineman, and Phyllis J. Walton. "Holy Hermaphrodites and Medical Facts: The Depictionof Hermaphrodites in Alchemy and Medicine." *Cauda Pavonis: Studies in Hermeticism*, n.s., 18 (Spring–Fall 1999): 32–36.

Walton, Michael T., and Phyllis J. Walton. "Being Up Front: The Frontispiece and the *Prisca* Tradition." *Cauda Pavonis: Studies in Hermeticism,* n.s., 17.1–2 (1998): 8–13.

Index